The Economics of Natural Gas Storage

Anna Cretì
Editor

The Economics of Natural Gas Storage

A European Perspective

Dr. Anna Cretì
Bocconi University and IEFE
(Centre of Research on Energy
and Environmental Economics and Policy)
via Roentgen, 1
20136 Milan
Italy
anna.creti@unibocconi.it

ISBN: 978-3-540-79406-6 e-ISBN: 978-3-540-79407-3
DOI: 10.1007/978-3-540-79407-3

Library of Congress Control Number: 2008940966

Cover design: WMXDesign GmbH, Heidelberg

Printed on acid-free paper

springer.com

Contents

Contributors

Edmond Baranes Faculté des Sciences Economiques, Université Montpellier 1, Espace Richter, av. de la Mer, CS 79606, 34960 Montpellier, France, edmond.baranes@univ-montp1.fr

Monica Bonacina IEFE (Centre of Research on Energy and Environmental Economics and Policy), Bocconi University, via Roentgen, 1, 20136 Milan, Italy, monica.bonacina@unibocconi.it

Alberto Cavaliere Università degli Studi di Pavia, Facoltà di Economia, Via San Felice, 5, 27100 Pavia, Italy, alberto.cavaliere@unipv.it

Anna Cretì IEFE (Centre of Research on Energy and Environmental Economics and Policy), and Bocconi University, Department of Economics, via Roentgen, 1, 20136 Milan, Italy, anna.creti@unibocconi.it

François Mirabel Faculté des Sciences Economiques, Université Montpellier 1, Espace Richter, av. de la Mer, CS 79606, 34960 Montpellier, France, francois.mirabel@univ-montp1.fr

Anne Neumann DIW (Deutsches Institut für Wirtschaftsforschung), Mohrenstraße 58, 10117 Berlin, Germany, ANeumann@diw.de

Jean-Christophe Poudou Faculté des Sciences Economiques, Université Montpellier 1, Espace Richter, av. de la Mer, CS 79606, 34960 Montpellier, France, jean-christophe.poudou@univ-montp1.fr

Antonio Sileo IEFE (Centre of Research on Energy and Environmental Economics and Policy), Bocconi University, via Roentgen, 1, 20136 Milan, Italy, antonio.sileo@unibocconi.it

Bertrand Villeneuve Université de Tours, CREST (Paris) and Laboratoire de Finance des Marchés d'Énergie, CREST – J320, 15 boulevard Gabriel Péri, 92245 Malakoff, France, bertrand.villeneuve@ensae.fr

Georg Zachmann LARSEN Laboratoire d'Analyse Economique des Réseaux et des Systèmes Energétiques, Université Paris, Sud 11, 27 avenue Lombart, 92260 Fontenay aux Roses, France, gzachmann@gmail.com

Foreword

I remember that the idea of this book emerged first in Toulouse, during the Third Conference on Energy Markets – 3 years ago now. Anna Cretì gave a talk on a model dealing with seasonal gas storage in the USA, and Christian Von Hirschausen was her discussant. Both of them were devoting their efforts to understand the natural gas market in Europe and the relevant liberalization process. I found their interest in storage rather original, so I encouraged Anna to collect the most original contributions on this topic.

Back in Milan with this idea in mind, she organized a working group at IEFE-Bocconi University, where she works. Then, during the following year, she exchanged ideas and organized several meetings with the book's contributors. She regularly invited the most important Italian gas sector representatives to these meetings, to make sure that the economic models were well suited to tackle the issues at stake in the European gas industry.

Now that the idea of this book has become real, I am very happy. The picture of the European gas industry that emerges from the collected work shows that there are still many issues to be solved before we reach the goal of a truly liberalized gas market. The models on storage and their applications to some very large gas consuming countries tell us that Europe is midway through the process. The institutional framework at the EC level is nearly completed, but its transposition into each Member Country is difficult. To make things worse, the dependency on foreign gas supplies is not expected to decrease.

I wish to congratulate Anna Cretì and all the book's contributors on their excellent work. I believe that this book really deserves to be read by energy (and non-energy) economists. I also recommend it to graduate students who like Industrial Organization applied models. Finally, practitioners and policy makers will appreciate the effort to adapt models to some very important European case studies.

Jean-Michel Glachant, Director of the Loyola de Palacio Programme on Energy,
Florence School of Regulation

Acknowledgements

There are many people without whose help this project would have never come to be. I wish to thank Jean-Michel Glachant and Christian Von Hirschausen for giving me the idea of writing a book on gas storage; Clara Poletti, Director of the Center for Research on Energy and Environmental Economics and Policy at Bocconi University, for trusting me when deciding to fund this project; my colleagues, the book's contributors, who agreed to work with me for more than 1 year, coming regularly to Milan; my other colleagues, who did not contribute to the book, but were patient enough to delay our meetings on different projects because I was working on gas storage.

If I look at the origin of my interest in natural gas storage, I have to thank the members of the Institut d'Economie Industrielle (IDEI), at the University of Toulouse, where several years ago I started working on energy. At that time, I was involved in a project on gas sector liberalization. The contacts with Gaz the France, and in particular with Corinne Chaton, gave me a wonderful opportunity to think about how Industrial Organization models can be used to understand modern gas industry. Bertrand Villeneuve worked with me enthusiastically in Toulouse, Milan and Paris to write our models. I owe the development of the most important insights on gas storage that I can now discuss in this book to Corinne and Bertrand.

In Milan, my young research team, Monica Bonacina and Antonio Sileo, worked very hard with me on the IEFE funded project "The Economics of Natural Gas Storage". This project has been very well managed also thanks to the help of IEFE's secretaries and administrative team: Lia Bertoglio, Daniela Cereda, Pinuccia Ganda and Dora Milanesi. I wish to thank all of them. I am also grateful to Angela Mezzanotte and Michel Roland who have patiently proof-read this book.

The Italian gas industry representatives contributed greatly to the development of my ideas. They have been very patient in discussing with me and the other authors of this book equations, results and contradictions between the theory and the real situation of the gas markets. In particular, Andrea Stegher and Gianluca Iannuzzi at Stogit and Valentina Infante at Edison Stoccaggio are to me the best examples of how firms which aggressively compete in the storage market are interested in cooperatively discussing economic models.

I know that my parents will be deeply moved on looking at this book. They live far from me, but they have always supported my decisions, even when they do not agree with me.

I think that this work will make my husband very happy and proud of me: he has never ceased to encourage me since we met during my Ph.D. He has fought against my pessimism and obstinacy, but he has always found the right way to tell me when I was wrong. If I am self-confident, it is because of his love.

Finally, this book is dedicated to my son, my first and unique fan. He is only a child now, later on he will understand.

Chapter 1
Gas Storage in Europe: Toward a Market-Oriented Approach

Anna Cretì

Storage is indispensable to the operation of the gas sector, since consumption, which is strongly influenced by weather, is seasonal and supply is relatively inflexible. Storing gas thus helps to avoid oversized extraction and transportation infrastructures, as well as to limit excessive price fluctuations.

Seasonal gas storage allows to inject gas during the summer, when demand is low, and to withdraw it during the winter, when demand increases. Given this pattern, storage capacity is usually measured by the amount of working gas that can be used throughout the year.[1] Storage also helps gas suppliers to face unpredictable demand fluctuations, including peak-days requirements exceeding the average winter consumption. Daily demand fluctuations can be dealt with storage inside pipelines, known as linepack and traditionally provided by the network operator.

Precautionary gas storage is used instead to manage the risk of supply disruptions, due to accidents or geopolitical reasons. The amount of storage devoted to the last purpose is also known as strategic storage and consists in gas stocks often managed under specific Government rules.

In all European countries, storage capacity was developed to cover the needs of a monopolistic market, and since then it has remained unchanged until the gas sector was opened to competition. In the current context of an evolving global market for natural gas and restructuring efforts in the European market, natural gas storage is gaining importance.

A. Cretì
IEFE (Centre of Research on Energy and Environmental Economics and Policy), and Bocconi University, Department of Economics, Via Roentgen, 1 20136, Milan, Italy
e-mail: anna.creti@unibocconi.it

[1] Due to pressure reasons, the amount of peak capacity guaranteed by storage companies decreases with the total amount of gas stored underground and is therefore typically larger at the beginning of the winter season and smaller towards the end of it, when inventories are lower. The volume of gas needed as a permanent inventory to maintain adequate reservoir pressures and deliverability rates throughout the withdrawal season is called cushion gas.

A. Cretì (ed.), *The Economics of Natural Gas Storage: A European Perspective*,
DOI: 10.1007/978-3-540-79407-3_1,

From a legislation viewpoint, in the first Gas Directive (98/30/EC) the regulatory regime introduced in the storage sector was unclear and lenient. To encourage market competition, in 2003 the second Gas Directive (2003/55/EC) clarified the originally ambiguous provisions for access to storage and introduced the concept of storage that is subject to third party access (henceforth TPA) or an open access regime. The recent third Energy Liberalization Package (2007) further introduces legal and functional unbundling to the TPA storage sector. With this trend, there is no doubt that storage in liberalized gas markets has some features beyond the traditional ones, such as seasonal balancing and security of supply.

In such a market-oriented approach, storing gas becomes an instrument for price arbitration. Price differential can be exploited in a tighter timeframe than the seasonal one (with daily, monthly, quarterly injections or withdrawals) or between the electricity and gas markets. In the latter case, if price soars at power exchanges, utilities can obtain rents from gas fired power plants by resorting to gas in storage instead of buying it in the spot market at a higher price, provided that the cost of access to storage is not too high. Moreover, in a decentralized gas system based on market exchanges, any utility with access to the transmission network (shipper) must fulfill commercial balancing obligations. Then utilities can usegas storage to avoid paying penalties for being unbalanced. Most importantly, storage can be used as a strategic tool in the imperfect competition dynamics that characterize the European gas sector, still dominated by former national monopolists.

The models on storage in the economic literature are not well suited to address the economic issues that have emerged in this sector in the aftermath of the gas market liberalization in Europe, as they were written before this path-breaking market reform. There is an extensive literature on storage (of any primary commodity) to smooth market prices; and on strategic stocks to face oil supply disruptions. I believe that these traditional models are worth summarizing, as they represent a well-established view on storage and form the necessary background to this book. I will survey them in the next Section and underline their weaknesses in addressing key issues on storage in the modern gas industry. In Sect. 1.2 I will argue that the complex set of storage usages in liberalized gas markets calls for a renewed theoretical framework. The objective of this book is to fill this gap.

1.1 Storage in the Economic Literature

The seminal models that have focused on the *role of stockpiling to stabilize prices* are the pioneering studies of Waugh (1944), Oi (1961) and Massel (1969). In a model where supply shifts occur with equal probability, Waugh showed that consumers are better off with fluctuating prices than if prices are stabilized to their simple arithmetic mean (corresponding to "full stabilization" policies). Using the same framework, Oi constructs a model in which a competitive firm earns greater total returns with unstable prices than with stable prices. The apparent discrepancy between the results of Waugh and Oi has inspired Massel to write a linear model in which the previous authors' results are shown to be special cases. Gains to producers and consumers individually may be positive or negative depending on demand/supply functions' steepness and on the relative variance of the shocks. However, those gaining form stabilization can compensate those losing, leaving everyone better off.

The pioneering works of Oi, Waugh and Massel have given rise to a literature whose main objective is to deduce the welfare implications of stabilization for producers, consumers and society as a whole. Welfare analysis of price stabilization has been extended to encompass alternative assumptions about price expectations, risk attitudes (Newbery and Stiglitz, 1981), and nonlinearity (Turnovsky, 1976). Costly inventories are studied by Helmberger and Weaver (1977) as well as by Edward and Hallwood (1980).[2]

The common approach of all these models is to compare market performance when storage does not exist, with performance when storage is used to stabilize price. It is worthwhile to stress that storage in this literature is made by a public authority, which is in charge of managing the buffer stock. The public authority sells at the stabilized price when there is excess demand and buys at the target price when there is excess supply, so the market clears. Hence the private storage industry is simply abstracted away. Another question not addressed by these models is whether stabilization schemes are optimal. All in all, the price stabilization literature shed little light on gas markets in Europe.

The second bunch of models studies *stockpiling as a device against disruption.* This idea has been mainly developed by the literature on exhaustible resources like oil. Many works have been motivated by the decision of the United States to develop strategic petroleum reserves in the aftermath of the OPEC oil embargo during the 1970s.

The oldest models argument a trade-off between current and future security of supply. In the event of a gas supply shock, if the scarcity value of domestic reserves is increased, providing an incentive for conservation to anticipate future emergencies, there is a motive for speeding up domestic production to reduce near-term economic losses if imports are interrupted, as long as supply cannot expand quickly in a disruption. The trade-off between current and future supply security has been subsequently analyzed by several authors (see for example Stiglitz, 1977; Sweeney, 1977; Wilman and Tolley, 1977; Hilmann and Van Long, 1983; Hughes, 1984). A typical analysis is to postulate a range of hypothetical supply interruptions for a representative year. Using a conventional description of supply and demand, a comparison of the pre- and post interruption markets reveals the changes in prices, payments for imports, and consumer surplus that made up the economic costs of interruption. Weighted by the probability of interruption, this comparative statics

[2] Helmberger and Weaver analyze different stabilization schemes in an intertemporal equilibrium model for a competitive market when costly inventories are held. A price stabilization policy forcing the market price to be higher than the competitive one will create excess storage and therefore will sacrifice economic efficiency. Producers gain from the government policy, while consumers lose. Edward and Hallwood consider a costly buffer stock whose objective is to maximize the joint expected benefits of the trading partners with respect to the intervention rules.

provides an estimate of the expected costs of supply insecurity. As stock draw down increases supply, the resulting reduction in the costs of interruption yields the estimate of the value of the reserve.[3]

However useful as these analyses might be, they ignored how to reach any desired stock level and how to deal with the uncertainty about the duration of the supply disruption. In the most significant attempt to address these questions, Teisberg (1981) develops a dynamic programming model that considers uncertainty in the duration of interruptions stockpile management rules, by minimizing a "US cost insecurity function" due to oil import disruption. It is often not desirable to use the entire stockpile during an emergency; part of the stockpile should be saved in case the supply interruption continues into the next period. With a very low stockpile, it may be best to build the stock even during a small interruption, as a hedge against the possibility of even greater losses during future periods.

Policy makers have rapidly recognized that in maintaining large stockpiles one country will absorb most of the direct costs while other countries will reap many benefits. Oil stockpiling can therefore be considered as a public good, and this lead naturally to a reluctance to build a large reserve (free-riding problem). In this perspective, the models on precautionary stockpiling have been generalized to include the analysis of strategic interaction among different importing countries on one side and exporters on the other. One of the first examples of this kind is the model by Nichols and Zeckhauser (1977). The authors argue that when stockpiling nonrenewable resources there are cases where stockpiling does not improve consumers' surplus. This result crucially depends on the way the interaction between producers and consumers' nations is modelled[4] and on the hypotheses on the probability of disruption.[5]

The literature on security of oil supply has not addressed the medium term problems that European countries face when the gas market is concerned. Either

---

[3] An interesting contribution to this debate is Lindsey (1989). He allows for repeated supply disruption of uncertain duration and arbitrary magnitude. With price rationing of world supply in a disruption, production during undisrupted periods should be speeded up unless costs are raising rapidly with cumulative extraction. With quantity rationing, production should also be speeded up unless the domestic economy is currently nearly self-sufficient.

[4] Crawford, Sobel, and Takahashi (1984) propose a dynamic bargaining framework in which the relationship between countries is a sequence of short-term negotiated agreements. The equilibrium involves the oil-rich country alone extracting for the first part of the relationship, exporting to smooth production in both countries until parity is reached. At that time, autarchy ensues, with the oil-rich and the oil-poor countries extracting until their stocks are exhausted. Total extraction is slower than the efficient path that would result if countries could organize their trading relationship by a single long term contract.

[5] In Devarajan and Weiner (1989), when disruption is expected to persist at the same intensity, each nation prefers the non-cooperative equilibrium to the cooperative one. If the disruption is expected to get worse, the noncooperative solution will lead to too little stockpile drawn down in the first period when compared to the cooperative solution. If several possible states in the oil market are possible (a normal market or a disruption of one or more possible sizes), the gain in free-riding to build oil stockpile are not worth the gamble and an aggressive stockpiling policy is preferable (Hogan, 1983).

the existing models ignore the existence of long-term contracts, therefore focusing on extraction rate of producer countries when there is a trade-off between present and future security, or they look at cartelized supply. Those are not the primary issues in managing secure gas services to Europe.

The most noticeable innovation, as for the last 20 years, in the storage literature is the introduction of highly nonlinear stochastic models, by Williams and Wright (1991). Most of their results are acquired through simulations, the stationarity assumptions facilitating the implementation of recursive methods. A number of valuable conclusions have been drawn, notably the clarification of the tension between short-run benefits (generally motivating public intervention, as in Wright and Williams, 1982) and long-run undesirable effects, which may dominate for producers and consumers. In other terms, the focus on stabilization is often inappropriately narrow.

Important developments of the literature on storage focus on the econometric methodology. Deaton and Laroque (1992) show that the simplest stochastic model, stationary and submitted to i.i.d. shocks, fails to reproduce the moments detectable in the data, notably the high level of prices autocorrelation.[6] Still, one possible explanation for some disappointing results is that the modelling of storage (costs, behavior of the speculators) may be oversimplistic or overoptimistic.

In more recent years, the gas sector liberalization has inspired a few empirical studies. has simulated the impacts of a strategic fuels reserve (SFR) designed to limit the increase in gasoline prices in the days following refinery disruptions. The core economic model of storage is inspired by the theory of commodity spot markets and inventories proposed by Pindyck (2001). When the economy experiences successive small disruptions, the simulations show that a SFR crowds out private storage and creates welfare losses. The author argues that "negative results are small, but their overall impact could be substantial given that small disruptions occur with much greater frequency". This calls for a more complete modeling framework, which includes the analysis of the frequency of the crisis. In the context of the European gas market, the crisis probability is in fact low, but the foreseeable economic consequences of a supply disruption quite dramatic.

Tacking advantage of the availability of high frequency data on spot and standardized futures market in the USA, Modjtahedi and Movassagh (2005) test the basic theory of decentralized storage based on the arbitrage theory, as in Pindyck (2001). On a similar vein, Uría and Williams (2007) show the importance of the temporal aggregation in a model that accounts for injection and withdrawal decisions as a function of the price spreads on NYMEX and the gas stock level. The authors suggest that, especially with monthly data, regulatory requirements and seasonal effects limit the responsiveness of injection decisions in California to the futures market. To date, it would be difficult to replicate this kind of analysis for the European gas market, which is quite immature. Therefore, unsurprisingly few works are devoted to "modern" storage in Europe, with the notable exception of Hoffler and Kluber (2007) and the research reports by CIEP (2006, 2008). The

[6] Deaton and Laroque (1996) and Chambers and Bailey (1996) have improved the adequation of the models to the data by reinforcing serial correlation in the shocks.

latter provide a general assessment of storage in Europe, while the first focuses on storage gas demand to the horizon of 2030. Both analyses emphasize the lack of investment in storage capacity at European level. This is one of the problems that this book investigates.

1.2 Storage in the Modern European Gas Industry

Gas storage in Europe has come to the center of the energy policy debate mainly due to very cold temperatures in winter 2005/06 and the price increases during that time period, especially the UK. This has brought attention to several limits to the operation of natural gas storage.

First, there seems to be no working market for commercial storage, except probably in the UK. As specified in the Eurostat report (2004) "Free gas [in Europe] (gas available on short and medium term basis excluded long term contracts) represents only 25 Gm^3 compared with the consumption representing 435 Gm^3 a year" In this context, storage is an additional adjustment tool which may even replace the spot market as regards satisfying the demand during peak periods. As a matter of fact, in most Countries storage is either a de facto monopoly or is characterized by imperfect competition.

Second, in continental Europe most countries are characterized by a lack of storage capacity with respect to flexibility needs of new entrants in the liberalized gas market. According to the ERGEG Report that monitored compliance with the guidelines set by European Energy regulators most of the top 15 Storage System Operators have declared no available capacity in 2005. Though such a result may also be due to a lack of transparency in the storage market, it is hard to think that capacity constraints will be removed in the next few years.

Third, from a regulatory perspective, European Directives let Member Countries opt between regulated or negotiated third party access to the extent that storage is not a natural monopoly and storage competition is feasible in principle. However, as a consequence of market imperfection in the operation of storage and limited capacity, even regulation of storage tariffs by an independent regulator may not be optimal.

Finally, as security of gas supply raises serious concerns, strategies against disruption are becoming of crucial importance in Europe. In 2005, one quarter of the EU primary energy consumption was based on natural gas, and imports from neighboring producers, mainly Russia, accounted for 46% of the total EU15 demand (Eurogas, 2006). The dependency on external supplies is going to worsen in the next years, as gas consumption in Europe is expected to grow whereas indigenous sources are forecasted to slow down. Including the new member countries, the European dependence rate for gas will amount to 50% in 2010, 62% in 2020 and 70% in 2030. The European Commission holds the position that liberalization improves security of supply and has reinforced this support in its recent policy related documents. It is not yet clear, however, whether precautionary storage against supply shocks must be set at each Member State level.

Difficulties in organizing market-oriented services, lack of capacity, access rules and imperfect competition, as well as storage and security of gas supply are the key issues addressed in this book. As in the implementation of the European Directives there is some degree of freedom at country level, thus some heterogeneity in the organization of gas services on a European scale, the aforementioned bottlenecks in the storage sector have different weights in different Member States. Therefore, each Chapter puts forth an innovative analytical model discussed or applied to realistically explain specific issues in one of the four big European gas consuming countries (France, Germany, Italy and the UK). Overall the book's collected works represent an original bridge between cutting edge economic analysis and technical as well as regulatory aspects of storage in Europe, ten years after the first Directive on gas sector liberalization.

1.3 Book Synopsis

1.3.1 Chapter 2

Within Europe, there are two different regimes under which natural gas storage is operated, depending on whether we consider UK or continental Europe. On the one hand there is a competitive market for (seasonal) natural gas storage in the UK benefiting from a functioning spot market for the commodity. As a result, operation of natural gas storage works similar to other functioning markets, hence quantities injected or withdrawn from facilities do not influence price formation on the spot market. On the other hand, storage facilities in continental Europe remain to be operated by market incumbents. Moreover, the only functioning spot market is Zeebrugge in Belgium, whose liquidity remain limited. With the majority of storage facilities operated following the "old" regime of monopoly behavior in the natural gas industry and a lack of competitive trading places, the use of natural gas storage is likely to follow a different pattern from the one we observe in the UK. In particular, quantities taken from the market and injected into storage potentially have a significant impact of prices at trading places, as well as additional quantities put into the market.

In Chap. 2, Neumann and Zachmann develop a simple framework determining the theoretical optimal use of natural gas storage facilities in which a trader maximizes profits. Storage is a function of spot and forward prices and costs, subject to the characteristics of natural closely facilities. This Chapter examines technical aspects of storage technologies as they are embedded in the empirical model tested on German data. Neumann and Zachmann investigate if storage operators/owners holding capacities in less developed markets differ in their use of natural gas capacities from companies active in a competitive market.

The main result of the empirical analysis is that the perfect arbitrage theory fails to explain storage operation in Germany, where the process of opening to competition is not that advanced. Possible explanations encompass not only technical limits to storage, but, most importantly, strategic behavior of market players active in natural gas storage in Europe, a topic that is analyzed in Chaps. 3 and 4 of the book.

1.3.2 Chapter 3

In a context where European gas markets are illiquid, developing spot markets, gas hubs and third party access to storage capacities could increase the liquidity of the supply and enable operators to discriminate between their gas supply sources in a shorter term. In particular, given that storage capacities have been transferred to third parties since August 2004, it is obvious that storage will become a significant flexibility tool when choosing a gas supply portfolio as it is observed in many American states.

Due to limited storage capacities, discriminating between storage and the spot market is not common in the European Union. Traditionally, storage is considered as a tool enabling to optimize the gas transmission system and to ensure continuity of the service. In Chap. 3, by Baranès, Mirabel and Poudou, storage not only has this public service dimension, but can also be used for influencing the strategic decisions made by competitors. This strategic dimension is reinforced by the fact that storage concerns an intermediate good and can therefore influence vertical relationships between oil-gas operators and suppliers through the spot market.

Baranès, Mirabel and Poudou analyze storage strategies when third party access to storage facilities is implemented. Access to storage facilities allows rival firms to adjust strategically the gas price on downstream market. Such a situation arises when the competitive suppliers are integrated with an upstream oil and gas company. When both producers and suppliers have market power, the spot price is influenced by storage decisions of suppliers: buying an additional unit in the first period and carrying it over into the next period instead of buying it in the second period pushes the first-period spot price up and the second-period spot price down. Since all rival suppliers buy at this spot price, they also benefit from price changes without incurring storage costs.

When taking into account the structure of the market, with imperfect competition in both production and supply, it a supplier owning a storage facility is not always interested in foreclosure. On the contrary, he might prefer to let his rival bear the costs associated with holding inventories, and benefit from the subsequent reduction of the spot market price. Baranès, Mirabel and Poudou discuss the likely effects of constraints such as seasonality or security of supply obligations on storage and competition in the gas market, thus providing a theoretical rationale to the issues empirically tested by the research carried out in Chap. 2. Chapter 3 also examines the related regulatory and market issues in the French gas storage system.

1.3.3 Chapter 4

Italy represents a good example of the impact of storage capacity constraints on gas sector liberalization: access to storage is rationed and the availability of sufficient storage capacity represents a barrier to entry in the market. Incumbents are less affected by capacity constraints because they both control the storage operator and dispose of a flexibility portfolio that enable them to substitute storage with other tools like import flexibility, indigenous production and interruptible contracts with industry. Therefore energy regulators forsaking liberalization goals should take proper account of that effect when regulating access to storage.

Though cost reflective storage tariffs are effective in controlling market power they are suboptimal both with respect to the allocation of existing storage capacity and to investment incentives. Regulated tariffs may be coupled with inefficient rationing procedures that allocate scarce storage independently of their value for gas sellers. Such a value is heterogeneous because it depends on the availability of storage substitutes whose distribution is asymmetric among gas sellers and typically gives an advantage to incumbents and to dual-fuel operators. A market mechanism based on a storage auction would be a better rationing tool in order to assign capacity according to the willingness to pay for it and improve the efficiency of storage allocation. Moreover the auction is supposed to assign scarcity rents to storage operators and can then work as an incentive to invest in new storage capacity. Relying on secondary market may improve storage allocation with respect to centrally planned procedures but scarcity rents may not be appropriated by storage investors.

Chapter 4, by Cavaliere, analyzes these issues in the framework of a dominant firm model that reflects the structure of the market for gas in most European Countries, under the hypothesis of a technological asymmetry to the extent that the competitive fringe is characterized by an higher (or even infinite) cost of storage substitute. The vertical relationship between access to storage and competition in the market for gas to compare market equilibrium will be considered in two cases. In one case access to storage is regulated with cost reflective tariff and storage capacity is rationed according to the share of each firm in the market for gas. In the second case storage is assigned through a multiple unit auction based on the marginal bid. After carrying out equilibrium analysis, the author compares social welfare and consumer surplus to find the most efficient rationing procedure. Therefore Cavaliere considers the strategic behavior that leads the incumbent to distort the auction. Welfare analysis may still establish that auctions are more efficient than the alternative procedure even accounting for strategic behaviour. The main implications of this model for the Italian gas storage system are discussed trough a detailed analysis.

1.3.4 Chapter 5

By diversifying the risk of disruption and financing pipeline construction, long-term contracts with producers are the primary supply instruments. Security of supply targets can also be met by increasing the system flexibility (interruptible contracts, cross-border pipeline capacity, liquid spot markets). However, these mechanisms have a limited capacity to absorb shocks that would endanger all the European countries at the same time (accident, civil war, terrorist attack). To ensure uninterrupted services in the short-medium term, precautionary gas storage is indispensable.

One of the milestones of the Security of Supply in the European Gas Market is the Council Directive 2004/67/EC of April 2004 concerning measures to safeguard security of natural gas supply. The Directive establishes a common framework within which the countries must define general security of supply policies and identifies a non-exhaustive set of instruments to enhance security of supply. Regarding gas storage, the Directive sets minimum targets, at national or industry level, asks for transparency of the storage policy and requires member states to publish regular reports on emergency mechanisms and the levels of gas in storage that the Commission will monitor – a procedure which to date is in place in the US only. Gas storage is, in fact, the primary flexibility mechanism that has to be used to absorb shocks that would endanger all the European countries at the same time. This kind of shock is persistent and independent from national policies.

Chapter 5, by Cretì and Villeneuve, focuses on the assessment of the optimal precautionary gas stocks that should be accumulated, knowing the potential minimum and maximum prices, the carrying costs and the probability of crisis. Most importantly, since the understanding of potential market failures or imperfections is of crucial importance in the perspective of the European Directive aimed at improving the security of gas supply, Cretì and Villeneuve analyze the effects of public interventions. The understanding of potential market failures or imperfections is of crucial importance in the perspective of the European Directive aimed at improving the security of gas supply. For example, stockholders may fear antispeculation measures taken once the crisis has occurred. This lack of protection of property rights could discourage storage completely; responsible policy could consist in a series of measures taken ex ante, as the authors argue.

Chapter 5 extensively refers to the UK debate on security of gas supply. Based on the relative advantages of imperfect policies in terms of total welfare, Cretì and Villeneuve clarify the potential incentives to store gas for security of supply in the most liberalized European gas market.

The Conclusions underline cross country aspects of storage services and focus on the main policy oriented messages that the models developed in this book deliver.

References

Chambers, M. J. and Bailey, R. E. (1996). A Theory of Commodity Price Fluctuations. *Journal of Political Economy, 104*(5), 924–957.

CIEP (2006). The European market for seasonal storage, Discussion paper.

CIEP (2008). Pricing natural gas: The outlook for the European market, Clingendael Energy Paper.

Crawford, V., Sobel, P. -J., and Takayaski, I. (1984). Bargaining, strategic reserves, and International trade in exhaustible resources. *American Journal of Agricultural Economics, 66*(4), 472–480.

Deaton, A. and Laroque, G. (1992). On the behaviour of commodity prices. *Review of Economic Studies, 59*, 1–23.

Deaton, A. and Laroque, G. (1996). Competitive storage and commodity price dynamics. *Journal of Political Economy, 104*(5), 897–923.

Devarajan, S. and Weiner, R. J. (1989). Dynamic policy coordination: stockpiling for energy security. *Journal of Environmental Economics and Management, 16*(1), 9–22.

Edward, R. and Hallwood C. P. (1980). The determination of optimum buffer stock intervention rules. *The Quarterly Journal of Economics, 94*, 156–166.

EUROGAS (2006). Annual Report.

Ford, A. (2004). Simulating the impacts of a strategic fuels reserve in California. *Energy Policy, 33*(4), 483–498.

Helmberger, P. and Weaver, R. (1977). Welfare implications of commodity storage under Uncertainty. *American Journal of Agricultural Economics, 59*(4), 639–651.

Hilman, A. L. and Van Long, N. (1983). Pricing and depletion of an exhaustible resource when there is anticipation of trade disruption. *The Quarterly Journal of Economics, 98*(2), 215–233.

Hogan, W. (1983). Oil stockpiling: Help thy neighbor. *Energy-Journal, 4*(3), 49–71.

Hoffler, F. and Kluber, M. (2007). Demand for storage of natural gas in northwestern Europe: Trends 2005–30. *Energy Policy, 35*, 5206–5219.

Hughes, H. A. J. (1984). Optimal stockpiling in a high-risk commodity market: The case of Copper. *Journal of Economic Dynamics and Control, 8*(2), 211–238.

Lindsey, R. (1989). Import disruptions, exhaustible resources, and intertemporal security of supply. *Canadian Journal of Economics, 22*(2), 340–363.

Massel, B. (1969). Price stabilization and welfare. *The Quarterly Journal of Economics, 83*, 284–298.

Modjtahed, B. and Movassagh, N. (2005). Natural-gas futures: Bias, predictive performance, and the theory of storage. *Energy Economics, 27*, 617–637.

Newbery, D. and Stiglitz, J. E. (1981). *The theory of commodity price stabilization*. Oxford: Oxford University Press.

Nichols, A. and Zeckhauser, R. (1977). Stockpiling strategies and cartel prices. *The Bell Journal of Economics, 8*(1), 66–96.

Oi, W. (1961). The desirability of price instability under perfect competition. *Econometrica, 29*, 58–64.

Pindyck, R. (2001). The dynamics of commodity spot and future markets: A primer. *The Energy Journal, 22*(3), 1–30.

Stiglitz, J. (1977). An economic analysis of the conservation of depletable natural resources. Draft Report, IEA, Section III.

Sweeney, J. (1977). Economics of depletable resources: Market forces and intertemporal bias. *Review of Economic Studies, 44*, 125–142.

Teisberg, T. J. (1981). A dynamic programming model of the U.S. strategic petroleum reserve. *The Bell Journal of Economics, 12*(2), 526–546.

Turnovsky, S. J. (1976). The distribution of welfare gains from price stabilization: The case of multiplicative disturbances. *International Economic Review, 17*, 133–148.

Urìa, R. and Williams, J. (2007). The supply of storage for natural gas in California. *The Energy Journal, 28*, 31–50.

Waugh, F. V. (1944). Does the consumer benefit from price instability? *The Quarterly Journal of Economics, 53*, 602–614.

Williams, J. C. and Wright, B. D. (1991). *Storage and commodity markets*. Cambridge: Cambridge University Press, New York and Melbourne.

Wilman, J. D. and Tolley, G. S. (1977). The Foreign dependence question. *The Journal of Political Economy, 85*(2), 323–347.

Wright, B. D. and Williams, J. C. (1982). The Roles of Public and Private Storage in Managing Oil Import Disruptions. *The Bell Journal of Economics, 13*, 341–353.

Chapter 2
Expected Vs. Observed Storage Usage: Limits to Intertemporal Arbitrage

Anne Neumann and Georg Zachmann

2.1 Introduction

Germany is one of Europe's largest natural gas importers and consumers. Given falling domestic reserves, natural gas storage therefore plays an increasingly important role. However, in regulatory terms Germany holds the "red lantern" in Europe, with very little institutional reform progress and a largely non-competitive natural gas sector. Subsequently, storage capacities are inefficiently used and the signals for new storage investment are distorted. In this chapter we analyze the structure of natural gas storage in Germany and apply a simple econometric model to see if a particular storage site is efficiently used. The chapter starts with some theoretical considerations about the theory of storage and provides some technical details of storing natural gas. We then introduce the natural gas storage activities in Germany (Sect. 2.3). It is dominated by depleted gas and oil fields, but aquifers and salt caverns also play a significant role. The inefficient access to existing storage sites of the incumbents has prompted the new market entrants to invest massively into new sites. In Sect. 2.4 we develop a model to evaluate the usage strategy of the observed use of storage with the "perfect arbitrage" solution. By comparing the optimal benchmarking behaviour with the real data, we can infer if the storage market works competitively. In Sect. 2.4 we apply the model to real-time data of a large storage site, Dötlingen owned by BEB. We find that injection and withdrawal decisions are not based on the profit maximizing behavior of a small player in a liquid market.

A. Neumann
DIW (Deutsches Institut für Wirtschaftsforschung), Mohrenstraße 58, 10117, Berlin, Germany
e-mail: ANeumann@diw.de

G. Zachmann
LARSEN (Laboratoire d'Analyse Economique des Réseaux et des Systèmes Energétiques), Université Paris, Sud 11, 27 avenue Lombart, 92260, Fontenay aux Roses, France
e-mail: gzachmann@gmail.com

A. Cretì (ed.), *The Economics of Natural Gas Storage: A European Perspective*,
DOI: 10.1007/978-3-540-79407-3_2,

2.2 Empirical Models and Theoretical Foundations

In general, storage transfers a commodity from one period to the next, including the related costs due to intertemporal arbitrage. Consumers holding inventories receive an income stream referred to as convenience becoming due in times of production and/or supply shocks. Therefore, theory implies a difference of spot and forward prices of a commodity at a level given by storage and interest costs (for storing) less convenience yields. Moreover, marginal convenience declines with increasing aggregate level of storage following a convex shape (Fama and French, 1987). The convex shape of the convenience yield implies a modest impact of changes in stock level on marginal costs of storage. Therefore, variations in spot prices are directly related to the benefit of holding inventory and inversely related to the correlation between spot and forward prices. Storage serves to balance short term differences in demand and supply. Entrepreneurial decision criterions for the use of storage are essentially described by: "Store until the expected gain on the last unit put into store just matches the current loss from buying – or not selling it – now" (Williams and Wright, 1991, p. 25). Storage facilities therefore induce arbitraging potential in functioning markets. Traders consider storage as an option derived as the sum of intrinsic and extrinsic values. In other words, the value derived from forward quotations and volatility of spot prices. Wright and Williams (1982) show that storage in a model where production is stochastic and both production and storage are performed by competitive profit-maximizers is favorable for consumers.

Deaton and Laroque (1996) investigate commodity prices for harvest assuming existence of speculators and competitive storage. Defining risk-neutral and profit-maximizing stockholders implies the nexus of spot prices over time periods. The authors show that the effect of storage on prices is only modest, but stronger on the mean and variance of the following period. Wright and Williams (1989) argue that backwardation[1] reflects a risk premium that drives futures prices down. Moreover, they argue that a negative price for storage is a positive difference between full carrying cost and expected rate of change of the spot price.

Markets for natural gas have been of interest for an application of storage theory. This is mainly due to the peculiarities of energy sources as compared to wheat or coffee: natural gas storage is limited by technical factors influencing operability of facilities induced by geological and technical characteristics, and strong seasonality. However, the existence of a number of spot markets (with futures and options traded at) for natural gas and the intertwining of former regionally segmented markets in the US resulted in applications of storage theory. Susmel and Thompson (1997) provide empirical evidence demonstrating that an increase in price volatility was followed by investment in additional storage facilities. The increase of the variance inherentes in spot prices (due to changing market structure and institutional framework) theoretically results in an increased use of storage (an increase in volatility increases the marginal benefit of holding inventory). An application

[1] Backwardation refers to a situation in which a commodity's future price for future delivery is below the price for immediate delivery.

to the Californian market for natural gas is provided by Uria and Williams (2007) arguing that injection decisions rather than resulting stock levels respond to price differences ("despite official seasons, regulatory requirements, and operational rigidities"). Using daily flow data the authors show that injection in Californian facilities increases slightly with a strengthening intertemporal spread on NYMEX. Serletis and Shahmoradi (2006) test the theoretical prediction that when inventory is high, large inventory responses to shocks imply roughly equal changes in spot and futures prices, whereas when inventory is low, smaller inventory responses to shocks imply larger changes in spot prices than in futures prices. Their tests on North American spot and futures natural gas prices confirm these predictions of the theory of storage. Wei and Zhen (2006), Dencerler, Khokher, and Simin (2005) and Khan, Khoker, and Simin (2005) model risk premiums and the dependence of futures prices on inventory levels with a focus on mean-reverting behaviour for natural gas among other US commodities. Chaton, Cretì and Villeneuve (in press) develop a model of seasonal natural gas demand taking into account the exhaustibility of the resource as well as supply and demand shocks. In a competitive setting the effect of policy instruments, i.e. tariffs or price caps, are investigated and applied to the North American market.

The technology of underground natural gas storage differs in the physical and economic characteristics of the sites. Deliverability rate, porosity, permeability, retention and capability of a site are the main physical of each storage type. To make operation of a storage site financially viable site preparation, maintenance costs, deliverability rates or cycling capacity are the main features. Keyes for profitable site operation are capacity and deliverability rate. The more natural gas injected or withdrawn the higher the economics of scale. Flexibility, and therefore the ability to react to short-term price signals, requires reasonable deliverability.

Depleted gas and oil fields (DGF, DOF) can be converted to storage while making use of existing wells, gathering systems, and pipeline connections. Natural aquifers are suitable for storage if the water bearing sedimentary rock formation is overlaid with an impermeable cap rock. Whereas aquifers are similar to depleted gas fields in their geology they require more base (cushion) gas and greater monitoring of withdrawal and injection performance. Deliverability of the site can be enhanced if there is an active water drive. The highest withdrawal and injection rates relative to their working gas capacity are provided by salt caverns. Moreover, base gas requirements are relatively low. Constructing salt cavern storage facilities in salt dome formations is more costly than depleted field. But the ability to perform several withdrawal and injection cycles each year reduces the per-unit cost of gas injected and withdrawn.

The fundamental characteristics of an underground storage facility distinguish between the characteristic of a facility (i.e. capacity), and the characteristic of the natural gas within the facility (i.e. inventory level). The total natural gas *storage capacity* is the maximum volume of natural gas that can be stored in an underground storage facility at a particular time. *Base gas* (or cushion gas) is the volume of natural gas intended as permanent inventory in a storage reservoir to maintain adequate pressure and deliverability rates throughout the withdrawal season. *Working gas capacity* is the volume in the reservoir above the level of *base gas* and is available to

Table 2.1 Storage costs

	DGF/DOF	Aquifer	Salt cavern
Specific investment costs per cm working gas [Euro per cm]	0.18–0.33	0.38–0.40	0.54–0.65
Specific investment costs in withdrawal rate [Euro per cm]	11.4–22.7	26.5–34.8	13.6
Total costs per cycled per cm working gas p.a. [Euro cent per cm][a]	5.86	6.73	9.81
Total costs per (cm/day) withdrawal rate p.a. [Euro cent (cm/day) a][b]	3.82	5.87	1.99

[a]Capital costs plus and variable operating costs
[b]Capital costs plus fix operating costs
Source: Following Grewe, 2005

the storage operator. *Deliverability* is a measure of the amount of natural gas that can be delivered (withdrawn) from a storage facility on a daily basis (often referred to also as deliverability rate, withdrawal rate, or withdrawal capacity). Deliverability varies and depends on factors such as the amount of natural gas in the reservoir, the pressure within the reservoir, compression capability available to the reservoir, the configuration and capabilities of surface facilities associated with the reservoir, and other factors. It is highest when the reservoir is full and declines as working gas is withdrawn. *Injection capacity* (or rate) is the complement of the deliverability and is the amount of natural gas that can be injected into a storage facility on a daily basis. It is inversely related to the total amount of natural gas in storage (EIA, 2004).

Depending on the type of storage, investment costs, lead times and operating costs differ. There are no exact figures on natural gas storage sites available, but Grewe (2005) provides some good estimates which are presented in Table 2.1.

2.3 Germany in the European Natural Gas System

Germany is a net-importer of natural gas. In addition to 12% domestic production, natural gas is imported mainly from four countries: Norway (12%), Netherlands (45%), Russia (34%) and UK (23%) (see Table 2.2). The role of natural gas storage is therefore essentially defined as to balance (seasonal and short-term) demand and to secure supplies in times of tight supplies. The geographic location in North–Western Europe with connections to the major transit pipelines and short-term trading places favors the use of natural gas storage facilities to benefit from short-term arbitraging possibilities.

In particular, increasing import dependency, not only during winter months, and the declining share of domestic production creates the necessity to use natural gas storage more extensively (cf. Fig. 2.1). The required investments into storage facilities and potential market-based usage of these sites is discussed later in this chapter.

Table 2.2 Natural gas in Germany (2007)

	Quantity (Bscm)	Share of total imports (%)
Domestic	14.30	
Netherlands	19.13	22.85
Norway	23.74	28.36
Russia[a]	37.95	45.33
UK	2.90	3.46

[a]Including other Europe and Eurasia

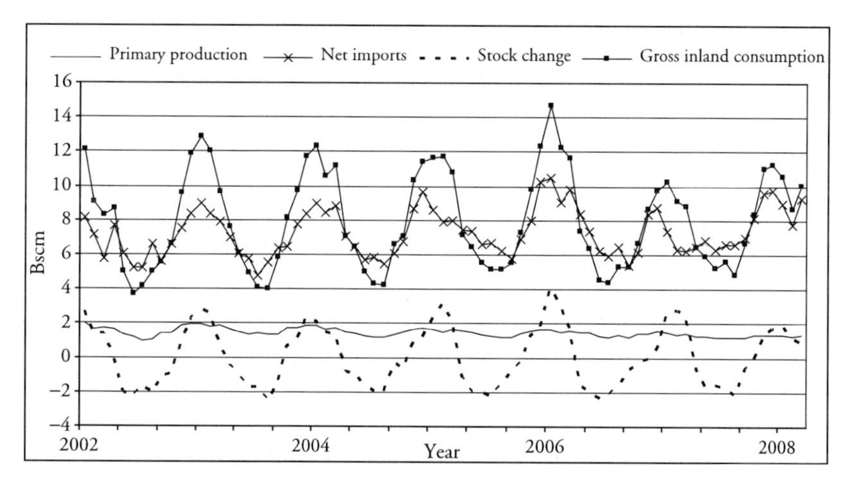

Fig. 2.1 Production, consumption and stock change in Germany: 2002–2008 Source: EUROSTAT

Germany accounts for 18 Bscm of working gas capacity and is therefore a major storage nation in Europe. The biggest European storage facility (in terms of daily peak withdrawal and injection rates) Epe is located in the western part of the country and is operated by 3 companies. Eight percent of the total working gas capacity (WGC) is located in aquifers which are geographically well spread over the country. Caverns provide 35% of total WGC and are mainly located in the North–Western, Eastern and Central (at the intersection of the MIDAL and STEGAL pipelines) parts of Germany. The main share of WGC (57%) is provided in depleted oil and natural gas fields. These storages sites are centred in Southern and spread in North–Western Germany. Few storage sites are located in central Germany and in South–Western Germany. The existing working gas volume in Germany has more than doubled since 1990 and provides a total working gas capacity of nearly the whole Dutch imports. Henceforth we provide a description of storage types in Germany and Fig. 2.2 shows the locations.

The first type of storage facilities addressed is aquifers. Total capacity of aquifers in Germany is 2.8 Bscm. The smallest facility contains 0.06 Bscm, the largest 0.63 Bscm and on average they have a total capacity of 0.35 Bscm. The total working gas capacity is 1.5 Bscm and thus around half of total capacity. Aquifer storage

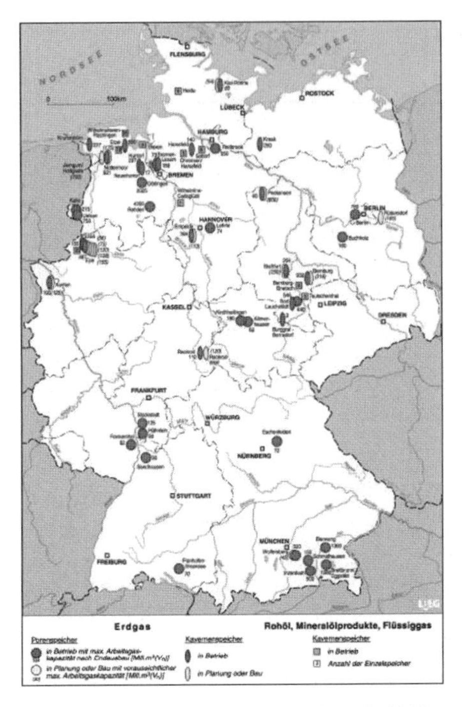

Fig. 2.2 Location of storage sites in Germany. *Source: Sedlacek, 2007*

sites are the smallest storage facilities with regard to WGC in Germany. The average Q_{max} injection rate is at $169.000\,cm/h$, the lowest is 45.000 cm/h, and the highest 400.000 cm/h. The biggest aquifer is located near Berlin. It has 0.78 Bscm working gas capacity and is owned by Berliner Gaswerke AG, a distribution company jointly owned by Gaz de France International S.A.S., Vattenfall Europe AG and Thüga AG. Verbundnetz Gas AG (VNG) owns another aquifer in this region (0.18 Bscm) which allows the company to store some of the imported natural gas from Russia. In the Western part of Germany, close to the Dutch border, RWE Netzservice GmbH operates a site which holds 0.22 Bscm working gas capacity. E.on Ruhrgas AG, Saar Ferngas AG and Gasversorgung Süddeutschland own an aquifer close to the river Rhine where advantageous geological conditions allow storing some 0.26 Bscm working gas capacity in total. Aquifers in South–West Germany near the cities Frankfurt/Main, Mannheim and Heidelberg are situated close to the pipeline-junction of the major transit pipelines MIDAL and SÜDAL.

Storage in depleted gas and oil fields (DGF, DOF) is less flexible. Total working gas capacity in Germany amounts to 10.9 Bscm. The smallest facility contains 0.4 Bscm, the largest 4.2 Bscm and on average they have a total capacity of 0.73 Bscm. Total capacity nearly doubles working gas capacity. Total available working gas capacity amounts to 2.6 Bscm. The average Q_{max} extraction rate is 313.000 cm/h, the lowest is 45.000 cm/h and the highest 1.200.000 cm/h. Most of the DGF are owned by natural gas importing companies, i.e. Wingas GmbH, BEB GmbH,

E.on Ruhrgas and RWE DEA AG. Most of the cavern storage sites are located in the Northwestern part of Germany in a large gas field which stretches from the North Sea and the Netherlands into Hamburg. Southern Germany, in the region around Munich and Rosenheim is home to five depleted gas storage sites with a total working gas capacity of 3.35 Bscm located in a large gas field which stretches from Vienna (Austria) to Munich.

Natural gas storage in (salt) caverns is the most flexible and allows frequent injection and extraction. Cavern storage requires less base gas which allows a higher share of working gas capacity. Total working gas capacity in Germany amounts to 6.7 Bscm. The smallest facility contains 0.02 Bscm, the largest 1.66 Bscm and on average they have a total capacity of 0.34 Bscm. Total capacity is 9.1 Bscm. The average Q_{max} extraction rate is 570.500 cm/h, the lowest is 100.000 cm/h, and the highest 2.450.000 cm/h. The ownership structure of cavern storage sites is more diversified than for aquifers or DGF. However, there remains a dominant position of market players such as E.on Ruhrgas AG (2.4 Bscm WGC), Verbundnetz Gas AG (1.4 Bscm WGC) and EWE AG (1.2 Bscm WGC).

German natural gas storage facilities are owned and operated by 22 storage operators. Wintershall (4.2 Bscm), E.on Ruhrgas (3.9 Bscm), BEB (2.5 Bscm), Verbundnetz (2.2 Bscm) and RWE DEA (1.9 Bscm) operate approximately three quarters of total WGC.

Considering different types of storage facilities shows a more diversified picture. Cavern storage facilities are owned by natural gas importing companies (E.on Ruhrgas AG and Verbundnetz Gas AG), regional transmission companies (EWE AG, Essent Energie Gasspeicher GmbH) and distribution companies (Stadtwerke Kiel AG, Kavernenspeicher Stassfurt GmbH, GHG Gasspeicher Hannover GmbH). However, the regional distribution companies are owned by market players such as E.on Ruhrgas AG, RWE Energy AG and Thyssengas. In essence, the German market for natural gas storage facilities is dominated by five market participants – incumbents from the “old” world.

Taken together, the location of storage sites in Germany is geographically well dispersed. Although most of the working gas capacity is located in North–West Germany, which is required given the German natural gas import structure and based on geographically favourable conditions, there appears to be significantly fewer facilities in the Ruhr region. The regions close to Stuttgart, Ulm and Southern Germany as well as around Dresden seemingly lack the possibility to balance demand at short notice. Planned projects in Germany are listed in Table 2.3 and are mainly investment in new cavern storage facilities, 15 of which are planned to become operational by 2015. The majority of these planned projects are located close to existing sites. In most cases the ownership structure stays the same to the already existing cavity. It is interesting to note that the only new participant is planning to construct a new site in Epe. The companies in the consortium are mainly municipalities and independent traders. However, overall these investments are undertaken by dominant market players. In total, expected working gas capacity is about 8 Bscm and therefore increases the current volume by nearly 45% (see Table 2.4).

Table 2.3 Key characteristics of German storage sites

Location	Operator	Type	Technical storage capacity (mcm)	Peak withdrawal capacity per day	Peak Injection capacity per day
Krummhörn		Salt Cavity	na		
Epe		Salt Cavity	1641	58.8	13.4
Hähnlein		Aquifer	80	2.4	1.4
Stockstadt	E.on Ruhrgas	Salt Cavity/Aquifer	135	3.3	2.2
Sandhausen		Aquifer	30	1.1	0.3
Bierwang		DGF	1360	28.8	13.2
Eschenfelden		Aquifer	72	3.1	0.8
Etzel	ConocoPhillips. E.ON Ruhrgas. StatoilHydro	Salt Cavity	560	31.4	12.96
Dötlingen		DGF	1076	13.44	12.96
Uelsen	BEB Speicher GmbH & Co. KG	DGF	520	5.88	5.88
Harsefeld		Salt Cavity	130	7.2	2.16
Rehden	Wingas	DGF	4200		
Kalle		Aquifer	215	9.6	4.8
Xanten		Salt Cavity	190	6.72	2.4
Nievenheim	RWE Energy	LNG Peak Shaving	14	2.4	0.11
Epe		Salt Cavity	414	12.48	4.08
Stassfurt	Kavernenspeicher Staßfurt GmbH	Salt Cavity	200	6	2.4
Buchholz		Aquifer	175	1.92	1.2
Bernburg		Salt Cavity	953	34.8	12
Bad Lauchstädt	VNG	Salt Cavity/DGF	1001	24.48	16
Kirchheiligen		DGF	190	3	3.36
Inzenham-West		DGF	500	7.2	3.36
Wolfersberg	RWE DEA	DGF	320	5.04	2.88
Breitbruno/Eggstätt	RWE DEA/Exxon Mobil/E.on Ruhrgas	DGF	1080	12.48	6
Peckensen	GdF Erdgasspeicher Deutschland	Salt Cavity	60	3	0.84
Huntorf		Salt Cavity	139		
Neuenhuntorf	EWE	Salt Cavity	17		
Nüttermoor		Salt Cavity	920		
Schmidthausen		DGF	150		
Lehrte	Deilmann-Haniel	DOF	40		
Reitbrook		DOF	350		
Fronhofen-Trigonodus	GdF Deutschland	Pore-Space	36	1.8	0.72
Bremen-Lesum	ExxonMobil	Salt Cavity	204	8.64	2.88
Frankenthal	Saar Ferngas	Aquifer	63		
Bremen-Lesum	Bremen Stadtwerke	Salt Cavity	78		
Berlin	Berliner Gaswerke	Aquifer	780		
Allmenhausen	Contigas	DGF	55		
Kiel-Rönne	Kiel Stadtwerke	Salt Cavity	60		
Kraak	Hamburger Stadtwerke	Salt Cavity	117		
Reckrod	Gas Union	Salt Cavity	82		
Epe	Deutsche Essen	Salt Cavity	181	0.4	0.2
		Total (in Bscm)	18388		

Source: Gas Storage Europe, 2008a and 2008b

Table 2.4 Expected investment into storage facilities

Location	Operator	Type	Investment	Expected WG capacity	Expected Date
Etzel	E.on Ruhrgas	Salt Cavity	New facility	2500	2013
Kiel-Ronne		Salt Cavity	New facility	50	2015
Etzel	EDF Trading/ EnBW	Salt Cavity	New facility	360	2011
Epe	Essent Energie Gasspeicher GmbH	Salt Cavity	New facility	200	2011
Epe 2A	Essent Energie Gasspeicher GmbH	Salt Cavity	Expansion	110	11/2008
Huntorf		Salt Cavity	New facility	150	2015
Nuentermoor	EWE	Salt Cavity	New facility	180	2015
Ruedersdorf		Salt Cavity	New facility	300	2015
Reckrod	Gas Union	Salt Cavity	New facility	30	2015
Anzing		Reservoir	New facility	165	2013
Berhringen		Reservoir	New facility	1,000	2013
Peckensen Phase 2	GDF Erdgasspeicher Deutschland	Salt Cavity	New facility	160	2010
Peckensen Phase 3		Salt Cavity	New facility	180	2014
Wielen		Reservoir	New facility	300	2014
Empelde	GHG	Salt Cavity	New facility	110	2015
Wolfersberg	RWE Dea	Reservoir	Expansion	45	2010
Xanten	RWE Energy	Salt Cavity	Expansion	125	2015
Frankenhal	Saar Ferngas	Aquifer	Expansion	130	2015
Epe	SPC Rheinische Epe Gasspeicher GmbH&Co KG/Essent Energy Productie B.V.	Salt Cavity	New facility	365	2010
Bemburg	VNG	Salt Cavity	New facility	300	2015
Jemgum	Wingas	Salt Cavity	New facility	1200	2015
Reckrod-Walf	Wintershall	Salt Cavity	New facility	120	2015
			Total (in Bscm)	8,080	

Source: Gas Storage Europe, 2008a and 2008b

German storage facilities are regulated according to EU legislation. The German Energy Law (Energiewirtschaftsgesetz, EnWG) transposes the Second Gas Directive 2003/55/EC into national law. It aims to provide a secure, reasonable priced and environmentally friendly supply of energy (Section 1 EnWG).

Section 28 EnWG requires access to storage facilities in the area of grid bounded supply of natural gas. Storage system operators have to provide other companies appropriate and non discriminatory access and supporting services if this access is technically and economically essential for an efficient grid access relating to the supply of customers (Section 28 (1) EnWG). However, storage system operators can refuse access if they prove that access is not possible due to operational or other reasons. Information on access conditions, storage facility location and available capacity has to be made available to interested parties.

In an accompanying ordinance, the Gasnetzzugangsverordnung (GasNZV) indicates that every interested party for using the distribution system shall be granted access to the grid and agreement must be submitted with the grid operator whose distribution system will be used for line entry or line exit (Section 3 (1) GasNZV). Section 15 GasNZV lays out the principles of storage capacity request and bookings. Grid operators have to publish a map covering the whole distribution system including all storage facility locations (Section 22 (1) GasNZV).

2.4 Market Based Use of Storage Capacities: A Model

The overview of German storage facilities and the corresponding operator in the previous section reveals a significant share of incumbents in the market for natural gas storage. In this section we test the hypothesis that the usage strategy observed at Dötlingen (a large depleted gas field operated by BEB) is not closely related to perfect or liquid market mechanisms. To evaluate the usage strategy of the facility, actual storage decisions have to be compared with some benchmark. Therefore, we proceed in three steps. First, we define the storage optimization strategies. Second, we calculate the behavior given the defined strategy (benchmark). Finally, we compare the benchmark behavior with the observed strategy.[2]

The benchmark that we want to compare with the observed storage decisions is the "perfect market" strategy. It is characterized by full price taking behavior of the storage customer. Therefore, the profit function can be written as $\Pi = \sum \Delta v_t p_t - c(\Delta v_t, p_t, V_t)$ where Δv_t is the storage decision, p_t is the price and $c(\Delta v_t, p_t, V_t)$ is the associated cost at time t.

In a second step, we calculate the storage customer's strategy. This is done by maximizing its profit with respect to stochastic prices, a non-linear cost function and non-linear constraints. Before presenting the algorithm the core components of the profit optimization are introduced.

A storage facility is essentially characterized by three factors: the injection rate, the withdrawal rate and the working gas volume (maximum less minimum volume).

[2] This section draws on previous work where we compare storage operation in the UK and Germany (Neumann and Zachmann, 2008). The basic idea is that a competitive market such as the UK will use natural gas storage according to the theory of storage.

We consider the maximum and minimum observed storage level as best proxy for the real upper and lower constraints. This approach has the advantage that not only the purely technical constraints are included but also non-technical obligations e.g. strategic reserves in case of bad weather, are incorporated.

Maximum injection and withdrawal rates are more difficult to deduce as they generally depend on the storage level. If, for example, a storage facility is close to its capacity limit it is technically more difficult to inject natural gas and if almost empty, withdrawal rates decline. Taking this behavior into account, we estimate the corresponding relationship using observed data. Therefore, we first extract the maximum injection and withdrawal speed for each storage level. Then we estimate the relationships between maximum injection rate and storage level, and between maximum withdrawal rate and storage level using a polynomial.

The cost function consists of four components: fuel cost, injection cost, withdrawal cost and storage cost. *Fuel cost* (fc) is a symmetric percentage (ϕ) of injections/withdrawals used for injection/withdrawal. Used fuel is valued at current prices and the fuel cost component is written as $fc_t = \phi \times \Delta v_t \times p_t$ (Δv_t is the storage decision). *Injection/withdrawal costs* (ic/wc) are additional cost depending only on injected/withdrawn volumes: $ic_t = \mu_i \times \Delta v_t$ if $\Delta v_t > 0$ and $wc_t = \mu_w \times \Delta v_t$ if $\Delta v_t < 0$. Finally, *storage cost* is the cost for holding gas in store: $sc = \varsigma \times V_t$. The assumptions for the four cost components are taken from Simmons (2000) and presented in Table 2.5.

To optimize its day-to-day injection/withdrawal decision, a storage customer needs to have some knowledge on future price developments. Futures and forward prices should represent the best guess of future spot price development, that can be represented by the so-called price forward curve (PFC). This PFC is calculated based on current futures prices. While weekly or monthly futures are traded near to spot month, seasonal or annual futures are traded for longer time horizons. Thus the PFC is calculated by smoothing and adding seasonalities (see Fig. 2.3).

Nevertheless it is clear to all market participants that future spot prices will generally deviate from the PFC. Therefore we assume natural gas spot prices to be stochastic in the short run while reverting to the corresponding PFC in the long run.

Table 2.5 Assumptions for the four cost components

Fuel used at each injection/withdrawal (θ)	1%
Cost associated to each injection (μ)	0.02 \$/MMBtu
Cost associated to each withdrawal (μ)	0.02 \$/MMBtu
Cost for holding natural gas in store (ζ)	0.40 \$/MMBtu

Source: Simmons, 2000

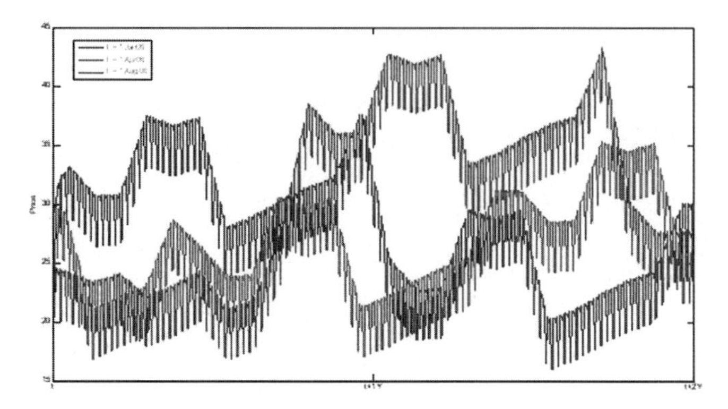

Fig. 2.3 Three price forward curves for TTF

The related parameters (mean reversion speed, volatility) are estimated using real data from the Dutch Title Transfer Facility (TTF).[3]

Different optimization algorithms for maximizing the profits from natural gas storage usage have been proposed in the literature. Generally two approaches can be distinguished. While solving a Bellman equation provides a closed form solution given certain price generating functions, Monte Carlo simulations are very flexible with respect to constraints and price models but have no analytic solution.

To cope with a nonstandard price function (reversion to moving mean) as well as nonlinearities in constraints and costs we follow Boogert and de Jong (2006) applying a Least Square Monte Carlo approach to natural gas storage contracts. Since identifying the optimal storage strategy is comparable to locating the exercise date of American options, Boogert and de Jong (2006) apply an option valuation algorithm proposed by Longstaff and Schwartz (2001). The general idea of the concept is to optimize storage usage decisions backwards in time using a discrete (daily) time grid, a discrete volume grid and n simulated price paths. The volume grid stretches from minimum to maximum storage level at equal distance volume steps: $Vol_{min} : VolStep : Vol_{max}$. These volume steps are defined to approximately represent a tenth of the daily decision spectrum (i.e. the difference of maximum injections and maximum withdrawals). Thus, at each day and volume combination, around ten different decisions are possible.

Time-values for a discrete set of allowed strategies are compared at each decision making point. Consequently, we first define a termination date and the payoff function at this date. We set the termination date $T = t_0 + 365$ (that is 1 year after the start date), and the payoff function at T is defined depending on the volume in storage at termination date (Vol_T). If the volume exceeds a desired level (Vol^*) the payoff is zero. We assume that a storage customer has to pay a punishment of double the time-value of the missing volume if the critical level Vol^* is undercut. This

[3] In an application to the UK market Hobæk et al. (2008) find a non-linear effect of storage on the relation of spot and futures prices.

yields:

$$Payoff_T^i(p_t^i, Vol_T) = \begin{cases} 0 & Vol_T \geq Vol^* \\ -2\,(Vol^* - Vol_T)\,p_T^i & \text{otherwise} \end{cases} \quad (2.1)$$

Therefore, the value of the storage contract $Value_T = Payoff_T(p_T, Vol_T)$ depends on the volume stored (Vol_T) and the price (p_T) at the termination date. The value is calculated for all simulated price paths and all allowed discrete volume levels. Departing from last day's storage values, we then move back a day and calculate the optimal decisions for all price paths and allowed volumes. We define the value in $T-1$ according to the current payoff of the optimal injection/withdrawal decision (*optDecision*), the discounted future value resulting from the volume after the optimal decision as well as the cost of the optimal decision:

$$\begin{gathered} Value_{T-1}^i(Vol_{T-1}, optDecision_{T-1}^i, p_{T-1}^i) = \\ [p_{T-1}^i \times optDecision_{T-1}^i + \delta \times Payoff_T^i(p_T^i, Vol_{T-1} + optDecision_{T-1}^i) + \\ -Cost(optDecision_T^i)] \end{gathered} \quad (2.2)$$

where δ is the discount factor.

The optimal decision, given each allowed (discrete) volume level, is derived by maximizing the current storage value, $Value_t^i(Vol_t, optDecision_t^i, p_t^i)$, with respect to the allowed discrete steps of $optDecision_t^i$. The volume level and the price path are the sole driver of the storage value and the $optDecision_t^i$ can thus be determined. According to (2.2) the value for each price-path volume-level combination is calculated. Similarly, the corresponding storage values for all points in time ($t = 1 : T-1$) can be determined:

$$\begin{aligned} Value_t^i(Vol_t, optDecision_t^i, p_t^i) &= p_t^i \times optDecision_t^i + \\ +\delta \times Value_{t+1}^i(p_{t+1}^i, Vol_{t+1}) &- Cost(optDecision_t^i) \end{aligned} \quad (2.3)$$

where δ is the discount factor.

To address the fact that storage customers can not know the exact spot price development in advance, n price paths are simulated according to the PFC and the estimated stochastic behavior. Furthermore, it has to be taken into account that the future storage value (that is needed to calculate the current optimal decision) is unknown to a storage customer as it depends on the future price development. Therefore, the methodology proposed by Boogert and de Jong (2006) implies a "Least Square Step" where, based on available information (current prices), the future storage value is estimated. The idea of this "Least Square Step" is to mimic the storage customers belief on the future storage value by regressing the future values, $Value_{t+1}^i(p_{t+1}^i, Vol_{t+1}^i)$, at each volume level and each price path on the current prices of this path p_{t+1}^i. The forecast of the future value is than given by $\widehat{V}_{t+1}^i = \widehat{\beta} \times p_t^i$.

The algorithm defined in (2.3) can now be iterated backwards from $t = T - 1$ to $t = 1$. The number of allowed storage decisions should be at least equal to three (injection at maximum capacity, withdrawal at maximum capacity, no operation) while a finer grid would allow for more precise results. As the quantity of storage levels is proportional to the resolution of the grid, the number of allowed levels is substantial. Furthermore, a high number of price paths is desirable to obtain reliable results. And finally, the observation period should at least contain one full cycle (365 days). Consequently, computation time becomes an issue and has to be carefully balanced with precision.

To make the benchmark strategy comparable with the observed strategy it is necessary that the benchmark strategy algorithm departs from the information set that was available to the actual storage customers. Therefore, the "ex post optimal strategy" (which implies perfect foresight of future prices) is a misleading benchmark for the observed strategy. Given imperfect price foresight (i.e. price simulation), the optimization has to be rerun for every point in time to assure that the price information are updated. Thus, the optimal strategy under imperfect price foresight is calculated in 365 subsequent rolling windows, each of which containing an updated 365 day price forecast ($t_{0,1} = 1:365, t_{0,2} = 2:366, ..., t_{0,365} = 365:730$).

Optimizing the strategy given imperfect foresight, the question arises which starting and end volume to assume for each run. Essentially, two "ex ante optimal strategies" are available: one departing from the past optimal decision, and one departing from past observed decisions. Starting each optimization from the observed level would on the one hand assure that each days information set is most accurately reflected. But on the other hand it implies that the cumulated decisions might surpass the technical constraints. Starting from the past optimal value, by contrast, assures that the cumulated decisions can be compared to the observed volumes. Since our analysis focuses on day-to-day injection/withdrawal decisions we optimize according to the observed initial volumes. Furthermore, final volumes are deduced from the observed data. As the optimization at the last day requires the end volume 365 days later, we require 730 days of observed storage volumes to obtain the optimal strategy for 365 days.

The Dötlingen storage site is operated by a significant market player: BEB is partly owned by ExxonMobil which is an important natural gas trader in Europe. Flow data for Dötlingen are available from October 2005 onwards. Table 2.6 summarizes the main characteristics of the storage site.

Table 2.6 Key characteristics of Dötlingen

Operator	BEB
Storage type	Depleted gas _eld
Used working capacity in GWh (technical max.)	8,847 (17,899)
Max. injection in GWh per day (technical max.)	109 (217)
Max. withdrawal in GWh per day (technical max.)	135 (217)
Available data	Daily aggregated injections and withdrawals

Source: Operator's website

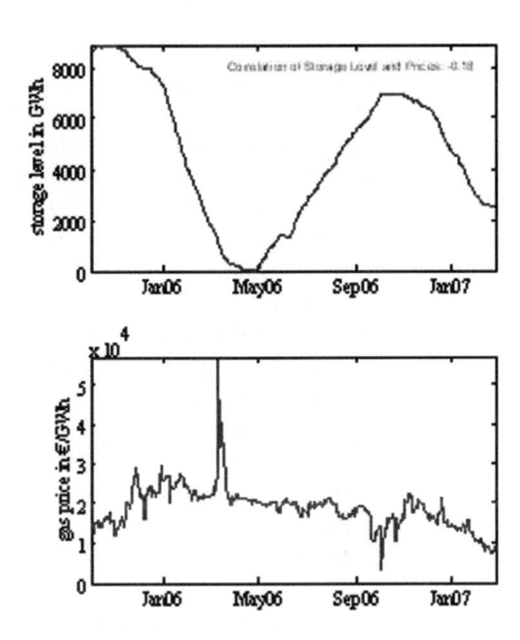

Fig. 2.4 Observed storage level and prices

Table 2.7 Correlations

Correlation of observed flows with benchmark case	0.08
For comparison	
Correlation of benchmark case flows with perfect foresight decisions	0.21
Correlation of observed flows with benchmark case at Rough (UK)	0.21

To understand the observed usage of the natural gas storage facilities we use prices at the Dutch gas exchange (TTF) which can be considered as reference and are shown for the time period under consideration in Fig. 2.4.

Applying the described algorithm provides the desired benchmark that can be compared to the observed data. The correlations of the observed storage flows with the corresponding benchmark (i.e. optimal decisions under imperfect price foresight) for the Dötlingen storage facility is 8%. This is a rather low degree of correlation when taking into account that the benchmark strategy is correlated with the perfect price foresight strategy at 21%. Even by international standards the explanatory power of the benchmark strategy for the observed flows is low since the equivalent for the Rough storage facility (UK) is three times bigger (23%). The low correlation of the observed flows and the benchmark indicate that natural gas injection and withdrawal decisions at Dötlingen are not based on the profit maximizing behavior of a small player in a liquid market (see Table 2.7).

Several explanations for this finding can be considered:

1. *Insufficiency of TTF.* Low liquidity of the spot and futures market as well as transport cost and potential congestion between TTF and Dötlingen might cause

prices at TTF not to be a good proxy for the true (but unknown) natural gas price at the Dötlingen entry/exit point.

2. *Technical considerations.* Storage operators could take into account additional technical constraints not considered in the benchmark strategy. For example, storage can serve to provide short-run balance of supply and demand, regulate the pipeline pressure, level injections/withdrawals in the system (e.g. LNG tanker arrivals).
3. *Strategic reserve.* Storage facilities might serve as a physical hedge and therefore be operated more smoothly than implied by pure arbitrage considerations.
4. *Cost of storage operations.* Optimization of the storage facility may not be based on the variable cost of storage operations. Storage contracts usually alloce a share of variable costs to flat-rate components.
5. *Exercise of market power.* Storage operators or customers can withhold natural gas in periods of low price elasticity. This creates potentials for strategic behavior for a sufficiently big player.

Given that possible explanations (2), (3), (4) and (5) apply to both the UK and the German market (although potentially at a different degree) our results show that a large part of the observed difference (of correlations) is due to the absence of a liquid and transparent German natural gas market. Therefore, the importance of a short-term trading and the role of natural gas storage are interrelated.

2.5 Conclusions

In this chapter we have shown that Germany is one of the biggest storage nations in Europe accounting for 18 Bscm of total storage capacity. Similar to other EU member states, indigenous natural gas production is declining whilst demand is rising. Therefore, dependence on natural gas imports is increasing, as well as the need to adjust to demand fluctuations both long and short-term. Therefore, the role of natural gas storage and a competitive usage of these facilities is gaining momentum in the process of market restructuring. Whereas most of the existing storage sites in Germany are owned by large companies active in long-distance pipeline transportation, investment in new capacities is also coming forward from market entrants. The development of a liquid trading point in the German pipeline system and the increasing interconnectivity with adjacent countries will further spur the development of new sites.

We have developed a model which uses real data for injection and withdrawal rates of a storage site in Germany, which is (1) favorable located, (2) owned by a big player (BEB), and (3) has published utilization rates. Furthermore, we use natural gas futures prices from the Dutch Title Transfer Facility to compare a competitive usage of the storage site with the observed behavior. The results show that the operation of this particular site in Germany does not follow the theory of storage. Even though there exists potential reasons to explain the discrepancy, our model also shows that German storage sites are not operated on a purely profit-maximizing behavior. This leads to the conclusion that the development of a competitive storage market ("merchant storage") is far from being completed. Hence, natural gas storage should be regulated and considered in a European context and the importance, both in economic and supply security terms, picked up by policy makers and other decision makers.

References

Boogert, A. and de Jong, C. (2006) Gas storage valuation using a Monte Carlo method, BWPEF 0704, December 2006, Birkbeck, University of London.

Chaton, C., Creti, A., and Villeneuve, B. (in press). Some economics of seasonal gas storage. *Energy Policy*.

Deaton, A. and Laroque, G. (1996). Competitive storage and commodity price dynamics. *Journal of Political Economy, 104*(5), 896–923.

Dencerler, C., Khokher, Z., and Simin, T. (2005). An empirical analysis of commodity convenience yields. Working Paper, June 2005, University of Ontario.

EIA (2004). The basics of underground Natural Gas Storage, Washington.

Fama, E. F. and French, K. R. (1987) Commodity futures prices: Some evidence on forecast power, premiums and the theory of storage. *Journal of Business, 60*(1), 55–74.

Gas Storage Europe (2008a). GSE available capacities. Brussels.

Gas Storage Europe (2008b). GSE storage investment database. Brussels.

Grewe, J. (2005). Auswirkungen der Liberalisierung auf die Erdgasspeicherung – Eine ökonomische Analyse für den deutschen Erdgasmarkt, Sonderpunkt Verlag Münster.

Hobæk, H., Lindqvist, O., and Løland, A. (2008). Risk premium in the UK natural gas forward market. *Energy Economics, 30*(5), 2420–2440.

Khan, S., Khokher, Z., and Simin, T. (2005). Inventories, convenience yields and mean reversion. Working Paper, October 2005, University of Western Ontario.

Longstaff, F. A. and Schwartz, E. S. (2001). Valuing American options by simulation: A simple least-squares approach. *Review of Financial Studies, 14*(1), 113–147.

Neumann, A. and Zachmann, G. (2008). Storage stories – operation of natural gas storage facilities in Germany and the UK. Mimeo, TU Dresden.

Sedlacek, R. (2007). Untertage-Gasspeicherung in Deutschland, Landesamt für Bergbau, Energie und Geologie, Hannover, Germany.

Serletis, A. and Shahmoradi, A. (2006). Futures trading and the storage of North American natural gas. *OPEC Review, 30*(1), 19–26.

Simmons and Company International Energy Industry Research (2000). *Underground Natural Gas Storage*. Houston, TX.

Susmel, R. and Thompson, A. (1997). Volatility, storage and convenience: Evidence from natural gas markets. *The Journal of Futures Markets, 17*(1), 17–43.

Uria, R. and Williams, J. (2007). The Supply-of-Storage for natural gas in California. *The Energy Journal, 28*(3), 31–50.

Wei, S. Z. C. and Zhen, Z. (2006). Commodity convenience yield and risk premium determination: The case of the US natural gas market. *Energy Economics, 28*(4), 523–535.

Williams, J. C. and Wright, B. D. (1991). *Storage and commodity markets*. Cambridge: Cambridge University Press.

Wright, B. D. and Williams, J. C. (1982). The economic role of commodity storage. *The Economic Journal, 92*(367), 596–614.

Wright, B. D. and Williams, J. C. (1989). A theory of negative prices for storage. *Journal of Futures Markets, 9*(1), 1–13.

Chapter 3
Natural Gas Storage and Market Power

Edmond Baranes, François Mirabel, and Jean-Christophe Poudou

3.1 Introduction

Retail energy market opening in Europe, cost reduction in electricity generation from gas, abundant gas resources available world-wide, contribute to explain natural gas expansion in Europe. Faced with this increase in gas needs in the European Union, authorities in Brussels underline the need for developing gas sourcing arrangements other than long-term contracts. Gas availability must be ensured over a shorter time scale in order to allow an actual "gas-gas" competition in the market.[1] The report published by the European Commission (2007) underlines the lack of liquidity of European gas markets. As stated in the DG Competition Report on Energy Sector Inquiry (10 January 2007, p. 8), there is "sub-optimal levels of liquidity in these [European] markets. In particular, the prevalence of long-term supply contracts between gas producers and incumbent importers makes it very difficult for new entrants to access gas on the upstream markets." In this context, developing spot markets, gas hubs and third party access to storage capacities may increase the supply liquidity and enable operators to trade off between their gas supply sources in a shorter term. Due to this flexibility, gas firms will be encouraged to implement an effective asset management for their supplies and diversify between available short and long-term sources. For instance in France, long-term contracts represent 80% of supply sources for Gaz de France. In the same way, long-term contracts represent 90% of supply sources for Distrigas in Belgium (annual report for 2005).

E. Baranes, F. Mirabel, and J.-C. Poudou
Faculté des Sciences Economiques, Université Montpellier 1, Espace Richter, av. de la Mer, CS 79606, 34960 Montpellier, France
e-mail: {edmond.baranes, francois.mirabel, jean-christophe.poudou}@univ-montp1.fr

[1] For an economic analysis of competition in gas market see Cremer and Laffont (2002).

A. Creti (ed.), *The Economics of Natural Gas Storage: A European Perspective*,
DOI: 10.1007/978-3-540-79407-3_3,

Supply source diversification is a flexibility tool for gas firms, particularly during peak demand periods and when resources are insufficient[2] with respect to the subscribed long-term contracts. In broad, a gas producer can manage its supply portfolio by using several flexibility tools: subscribing long-term contracts, upstream vertical integration,[3] or even using spot markets for obtaining the quantities which are not covered by the internal supply sources and long-term contracts. Lastly, the operator can use storage capacities to satisfy the demand increase during peak periods. Following the European Union's decision, storage capacities have been transferred to third parties since August 2004 (third party access to storage). In this context, it is obvious that storage will become a significant flexibility tool when choosing a gas supply portfolio as it is observed in many American states. This should enable gas operators, who are already in the market, to use another short-term adjustment tool in addition to the spot market.[4]

It appears that gas storage incorporates a strategic dimension additionally to traditional functions, as optimization of the gas transmission system and service continuity. We focus here on this strategic dimension of storage in gas competitive markets and analyze how it allows firms to exert market power.

This Chapter is organized as follows. As it will be argued in Sect. 5.2, economic literature has addressed the issue of market power and storage. In order to analyze further some of the relevant questions concerning gas storage and market power, Sect. 3.3 is devoted to the development of a basic but specific model of industrial organization of gas markets, which explains gas storage affects competition in gas markets between symmetric firms. In Sect. 3.4, we explore wether a priority access to storage facilities (creating a leadership) modifies playing field between actors. In Sect. 3.5 we tackle the important issue of legal and functional unbundling of storage activities, which is in debate in Europe and in the United States. In Sect. 3.6, we consider more precisely storage facilities which can be used by operators introducing two alternative storage facilities that have different degree of flexibility (e.g. salt caverns vs. depleted fields or aquifer caverns) related to injection/withdrawal scheduling. The last part of the Chapter (Sect. 3.7) is devoted to the analysis of the French natural gas storage sector. Proofs appear in the Appendix.

3.2 The Issue of Gas Storage and Market Power in the Literature

Economic literature on storage activities is relatively extensive. Traditionally, storage is considered as an investment enabling firms to adjust their supply when demand is uncertain or exposed to cyclical fluctuations.

---

[2] As specified in the Eurostaf report (2004) "Free gas [in Europe] (gas available on short and medium term basis excluded long-term contracts) represents only 25 Gcm compared with the consumption representing 435 Gcm a year."

[3] For instance, this has been done by merger and acquisition of holdings in oil and gas companies.

[4] Since 1998 in Europe, gas hubs have been established. For example, these spots markets are located in Bacton (UK), Zeebrugge (NL) and Emdem (Germany).

Three main motivations are identified in economic literature for explaining the benefits of storage for firms: speculation, precaution and seasonal production smoothing. First, the *storage speculative* function is relatively well accepted. In this case, storage enables firms to obtain a positive income faced with an exogenous shock which, for instance, influences the market price of the stored good. Second, the *precautionary motive* is a regulatory function; the stock allows firms to regulate market supply in answer to an uncertain demand when their production capacity is not very elastic. The analysis of this motive is developed in Chap. 4 of this book. Lastly, firms may choose storage in order to *smooth the cyclical fluctuations* of the demand. In addition to these three traditional functions, storage is a major subject in literature about oligopolistic competition within a dynamic context. Thus, for instance, Kirman and Sobel (1974), Philps and Richard (1989) study storage in a context of intertemporal price discrimination. In this case, storage introduces an intertemporal fixed price reliance in as much that decisions made during a certain period depend on actions from previous periods.

Another function has been identified in the economic literature: a *strategic function of storage*. This has initially been analyzed by Arvan (1985), Saloner (1987) and Pal (1991, 1996). Storage ensures a strategic function as it influences the future decisions of rival firms. Indeed it may be used by firms as a commitment means based on quantities. An oligopolistic firm may be induced to invest in storage capacities to preempt the future production of its competitors. In line with this perspective, Saloner (1987) and Pal (1991, 1996) consider a duopoly model in which, firms choose their advance production (which is assimilated to their storage level) during the first period and then, over a second period, sell their products in the market. With a Stackelberg leadership, they show firms may be induced to produce in advance even if their production is more expensive during the first period. Poddar and Sasaki (2002) examine incentives for firms to produce in advance in a multiperiod competitive setting. They show that advance production can be a strategy to create endogenously a Stackelberg leadership. As advance production, gas storage might be a tool to implement such a strategy.

A recent trend of the economic literature focuses on *gas storage, market power and regulation*. Esnault (2003) studies the need for storage of suppliers on deregulated gas markets. He shows that, in France and importing countries, as storage is a scarce resource, the Third Party Access must be completed by a regulation on reservoirs in order to introduce effective competition on the gas scene and to create efficient trading places. In Baranes, Mirabel and Poudou (2005), it is shown that storage facilities can be strategically used as a foreclosure tool when Third Party Access is regulated.[5] Breton and Kharbach (2008) consider the use of access-to-gas storage in a seasonal model. In a duopoly setting, they find that welfare is higher under vertical integration with an open access system than under separate management of storage. In this Chapter we model a stylized organization of vertically related gas markets and we can show that an upstream leadership in the access to storage facilities leads the dominant firm to adopt a strategic storage decision. This strategy

[5] We will come back to these arguments in Sect. 3.5.

consists in stockpiling more than supplied in the downstream market. This behavior is a part of a raising rival's cost strategy for the leader.

An interesting analysis on gas storage and market power is the one of Durand-Viel (2007) which studies the effect of storage decisions on upstream resource prices. In a two-tier oligopolistic structure, it is shown that, storage allows suppliers not only to preempt future demand, but also to counter upstream producers' market power. Indeed, the traditional vision of third-party access supposes that incumbents have incentives to deter entrants from storage capacities (see Sect. 3.4 for this argument). However, when taking into account specificities of gas market structure, it is shown that a storage facility owner is not always interested in foreclosure. He might prefer to let his rival bear the costs associated with holding inventories, and benefit from reductions of the spot market price.

From an empirical point of view, a recent academic literature focuses on the links between storage levels and natural gas prices. It is underlined that storage has an impact on the volatility and level of natural gas prices. For example. Mu (2007) shows that publications of statistics on storage levels in US is an important determinant of natural gas price volatility. Moreover, using Energy Information Administration data from March 1993 through June 2004, Modjtahedia and Movassagh (2005) show that the level of natural-gas working storage had a statistically significant effect on the basis of spot prices. In another very interesting paper, Egging and Gabriel (2006) study the impact of storage capacities on market power in the European gas market. Modelling four market scenarios to gauge the effects of market power, increased pipeline capacity to the United Kingdom, as well as the importance of storage, they find that ample storage capacities are key to favorable market conditions for consumers.

With regard to theoretical and empirical literature just mentioned, our model integrates a relationship between storage decisions and intermediate price within the spot market. In the framework we develop in this Chapter, dynamics of the spot market integrate externalities related to strategic storage decisions; this contributes to increase the gap between the gas price on spot market and the corresponding marginal cost.

3.3 A Basic Duopoly Model

In the following we present a model[6] of an industry with 2 operators competing in the natural gas market bearing an exogenous marginal supply cost $\gamma \geq 0$ (i.e. international price or long-term contracts). Consumers address demand to downstream firms, which is denoted by $p(Q)$ where Q is the aggregated volume of gas traded in the market. We assume that final gas demand is expressed as the normalized form $p(Q) = 1 - Q$. We focus here on the demand during peak demand periods (winter season) which requires operators to use storage facilities. We denote by q_i, the

[6] A more general version of this model can be found in Baranes, Mirabel, and Poudou (2007).

quantity of gas offered by a firm $i = 1,2$ in the downstream market. In order to provide this offer, firm i may use a storage facility operated at unit cost c by an independent storage firm. Therefore y_i represents volume injected[7] (during low-demand periods i.e. summer) in a storage facility by firm i, paying a unit price $a \geq c$ for access to storage facilities.

Two cases may be encountered according to whether the quantity of stored gas is sufficient or not to supply the downstream quantity of gas, q_i. Gas supplier may trade on the spot market for buying or selling gas in quantity z_i. It should be noted that z_i can be positive or negative trades on spot markets. When z_i is positive, this means that firm i has a diversified portfolio since it uses both spot market and storage facility to secure its supply. Conversely (when $z_i < 0$), firm i withdraws gas from the storage facility to sell it on the spot market. We refer to the latter situation as *strategic storage*. In this analysis, storage and spot can be considered as both substitutable or complement supply processes. Therefore, for each gas firm, there is a relationship between his supply in the final market, the quantity of gas injected in the stock and the position in the spot market, this leads to the following relation:

$$q_i \leq z_i + y_i. \tag{3.1}$$

We denote by s the spot price of gas, which therefore depends on volumes bought and sold in this market by gas firms and traders. In the spot market, 2 (pure) traders – indexed by j – compete with gas firms; a trader's (net) supply is denoted w_j. Actually, pure traders do not exist in gas hubs, they are operators trading gas on other geographical markets. Here the assumption is made that these firms are operating on the spot market as arbitragers taking advantage of price variations of the energy resource. A trader has a seller position when $w_j > 0$ and a buyer position whenever $w_j < 0$. We consider that these traders buy gas in similar cost conditions as gas firms (i.e. at cost γ).

Focusing on market power, we assume firms and traders maximize their profits which are given by:

$$\Pi_i(q_i, q_{-i}, y_i, y_{-i}) = p(Q)\,q_i - (a+\gamma)\,y_i - sz_i \quad \text{for a gas firm } i \tag{3.2}$$

$$\pi_s = (a-c)(y_1 + y_2) \quad \text{for the storage firm} \tag{3.3}$$

$$\pi_j(w_j, w_{-j}) = (s-\gamma)\,w_j \quad \text{for a trader } j \tag{3.4}$$

Figure 3.1 depicts the industrial structure which is considered in this model.

We solve a three-stage game to analyze strategic decisions concerning gas storage.[8] In the first stage, gas firms determine their storage strategy (i.e. they choose y_i).

[7] Indeed, these volumes are planned to be withdrawn in further periods.

[8] Even if this game is not dynamic, it is sequential. So to fix ideas, we can identify the first stage as the low demand period (indeed normalized at a zero level) and the second and the third stages as peak demand periods.

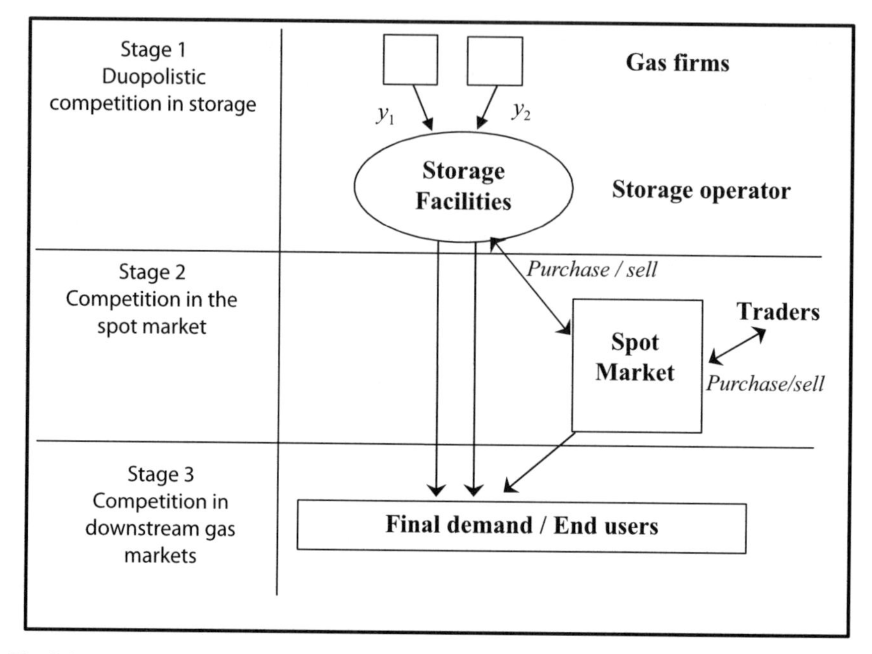

Fig. 3.1 Industrial structure in the basic model

In the second stage, gas firms and traders determine which purchases and sales are to be performed in the spot market (choosing w_j and z_i). In the third stage, gas firms compete in the downstream gas market (choosing q_i). We look for subgame perfect equilibria of this game. For simplicity, we assume that γ is equal to a given gas production cost which is normalized to zero.

The *third stage* consists in determining a traditional Cournot symmetric equilibrium[9] on downstream market in which firms bear a supply cost equal to the market spot price, s. In the following, we note $q^*(s) = \frac{1-s}{3}$ the equilibrium quantity for each firm $i = 1, 2$.

At the *second stage*, according to their observed positions in storage y_i, firms trade quantities $z_i = q_i - y_i$ in the spot market while traders adopt a position w_j. As a consequence, spot market clearing allows to determine the equilibrium price s such that:

$$q_1^* + q_2^* - (y_1 + y_2) = w_1 + w_2 \Leftrightarrow 2q^*(s) = y_1 + y_2 + w_1 + w_2. \tag{3.5}$$

Hence, the inverse demand in the spot market can be expressed as:

$$s^*(\cdot) = S\left(\frac{y_1 + y_2 + w_1 + w_2}{2}\right) \tag{3.6}$$

[9] This assumption is relevant since natural gas is a homogenous good.

where $S(x) = q^{*-1}(x)$. Using (3.6) and maximizing the profit in (3.4) we get the symmetric Nash equilibrium in the spot market, $w^*(y_1+y_2) = \frac{2}{9} - \frac{1}{3}(y_1+y_2)$. Then from (3.6), we deduce the equilibrium price of the spot market as a function of storage strategies:

$$s^*(y_1+y_2) = \frac{1}{3} - \frac{1}{2}(y_1+y_2). \tag{3.7}$$

Finally, in the *first stage*, gas firms choose which quantity of gas they want to store at the cost a in order to sell it in the downstream market or in the spot market. Maximizing profit in (3.2) for each firm, gives the following first-order condition, where y^* is the individual storage decision:

$$\underbrace{p'(\cdot)q^*(s^*)q^{*\prime}(s^*)\,s^{*\prime}(2y^*)}_{\text{final market effect}} - \underbrace{[q^*(s^*)-y^*]\,s^{*\prime}(2y^*)}_{\text{spot market effect}} = \underbrace{a-s^*}_{\text{cost effect}} \tag{3.8}$$

From (3.8), we can identify three various effects of any variation of stored quantities, y_i, on the profit of a gas firm: a final market effect, a spot market effect and a cost effect. The first effect (final market effect) is negative and measures how an increase of stored quantities decreases the profit for downstream firms. The intuition is that an increase of stored quantities yields a decrease in the spot price which in turns reduces the sourcing cost. This leads to an increase in the rival's total supply in the final market and to a decrease in the final price. The second effect represents the spot market effect for which an increase in stored quantities results in a decrease in the spot price. The intensity of this effect depends on gas firms' positions, $q^*(s^*) - y^*$, in the spot market. Finally the last effect is a cost effect which measures the net marginal cost of storage.

Solving the expression (3.8) gives the storage equilibrium,[10] $y^* = \frac{2}{75}(11-27a)$, and a resulting spot price equals $s^* = \frac{1}{25}(1+18a)$. From a general point of view, it can be seen that gas volumes are stored at the equilibrium so it is profitable for all gas firms as a commitment strategy (for not too large values of a). Moreover even if traders are worse off than without storage, it can be shown that there exists values of the access price for which storage improves social welfare.

From the equilibrium of this basic model, we can derive benchmark results on the interplay between competition in gas market and storage strategies.

Proposition 1. *At the symmetric subgame perfect equilibrium, no strategic storage occurs. Gas firms secure their supplies using the spot market in order to serve the downstream gas demand. This result applies even if the equilibrium spot price is higher than the access price to the storage facility.*

This result means that gas firms store insufficient quantities of gas compared to their supply in the final market. Indeed at the equilibrium, to adjust their sourcing operations they may use both spot market and storage capacities as flexibility tools. However, it is necessary for firms to use the spot market in order to complete their

[10] Remark that if $a > \frac{11}{27}$ operators source their gas only from the spot market since $y^* = 0$.

gas sourcing. It should be noted that even though spot price is relatively high,[11] the equilibrium strategy employed by gas firms is not an exclusive adjustment of their sourcing from the gas storage. However, the intuition of the result is straightforward: when firms choose an exclusive supply from gas storage, spot price turns equal to zero as demands in the spot market collapse. Therefore, in these conditions, firms have incentives to deviate and to resort to spot market.

3.4 Preemptive Access to Storage Facilities

In several countries in Europe, historical incumbents benefit from a preemptive access to storage facilities mainly because they are involved in gas storage operations. Typically, this configuration corresponds to the French situation which will be depicted in Sect. 3.7.

From a general point of view, preemptive access to storage facilities can be considered as a kind of leadership in storage decisions. As our basic model is concerned, taking into account this preemptive access assumption is made introducing a sequential move in the first stage of the game (i.e. storage decisions). In the duopolistic framework we consider, it implies that one of the two firms acts as a Stackelberg leader in the storage first stage. Arbitrarily we denote by $i = 1$, the leader firm and by $i = 2$ the follower. As a matter of fact, Proposition 1 in Sect. 3.3 depicts the equilibrium situation when access to storage facilities is absolutely fair and non discriminatory (no leadership).

To determine the subgame perfect equilibrium of this game with storage leadership, we just have to reconsider the equilibrium of the first (storage) subgame taking as given relations (3.6) and (3.7) from Sect. 3.3. From this leadership equilibrium, we can derive the following result.

Proposition 2. *The leader raises its rivals' cost using strategic storage whenever the access price is sufficiently low. Moreover strategic storage occurs also when the regulated access price is cost oriented.*

According to this result, storage can be a tool to assess market power for gas firms. Proposition 2 can be illustrated as in Fig. 3.2.

In the left part of the figure, the storage access price is relatively small, and we see that the leader decides to store significant volumes (i.e. $\hat{y}_1 > \hat{q}_1$). In this case, the follower (or the fringe of following firms) reacts by purchasing part of its sourcing operations in the spot market (i.e. $\hat{q}_2 > \hat{y}_2$), which turns out to be relatively more expensive at the equilibrium. Similarly, when the storage access price increases (right part of the figure), the leader reduces his stored quantities and he even interrupts storing activities and only uses the spot market for very high values of a. This induces the follower to react using a relatively more expensive storage strategy. From

[11] More precisely when $s^* \geq a$, it is the case whenever $a \leq \frac{1}{7}$.

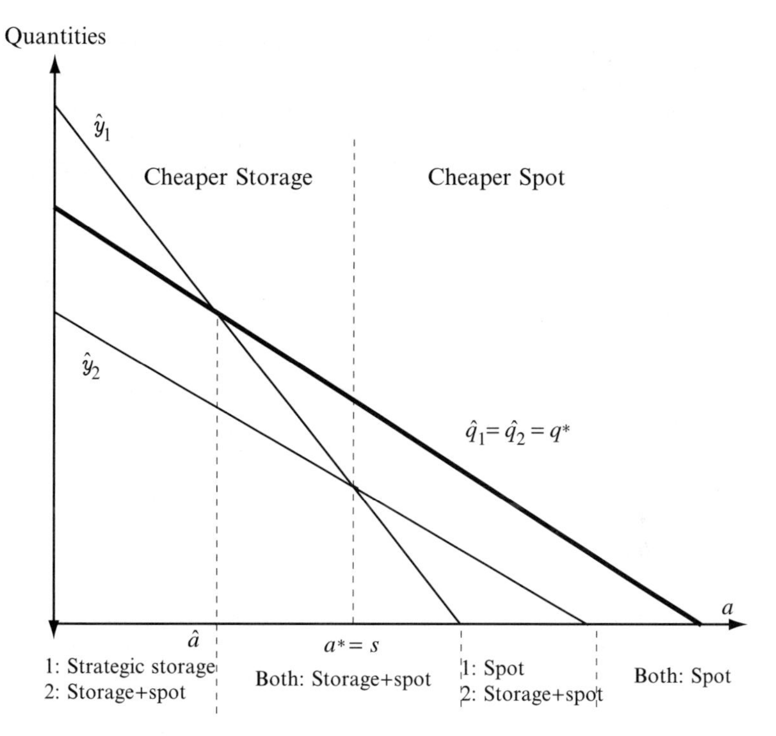

Fig. 3.2 Preemptive access and strategic storage

the leader point of view, this behavior can be seen as a part of a "raising rivals' cost" strategy.[12] For relatively low access price levels, the leader preempts storage facilities in order to sell excess quantities in the spot market. As a result, the spot price decreases and the follower is pushed toward the spot market to complete its gas sourcing even if storage is less costly. When the access price is higher, this effect remains unchanged but in a less intuitive manner. If storage capacities were actually restricted,[13] the follower would prefer to supply smaller volumes on the downstream market rather than using spot market in order to satisfy the demand. This can be explained by the strategic position of the leader in the spot market: using the spot market would be too costly for the follower.

Although this strategy may seem very profitable for the dominant firm, it may have a negative impact at a social level: decreasing profits of follower, storage owner and traders, and lowering consumers' surplus. It then appears necessary to study impacts on welfare[14] when this type of strategy is applied. Proposition 2 states that

[12] As defined by Salop and Scheffman (1983).

[13] Notice that when storage facilities are capacity constrained (as it is actually expected), the strategic storage behavior is all the more harmful for fringe firms.

[14] Social welfare is defined as the sum of the consumers' surplus and profits of gas suppliers, traders and the owner of the storage facility.

a social welfare-maximising access price (for example as fixed by a regulator), will not prevent strategic storage. This is actually the case when the optimal access price lies in the range where strategic storage represents an optimal strategy for the leader. This means the regulator prefers such a strategy to maximize welfare. For relatively low storage cost values, the socially optimal access price is cost-based and lower than the value below which the leader chooses a strategic storage behavior.

In this case, it is socially preferable to let the leader preempt capacities in the storage facility. On the one hand, stored quantities in excess are sold by the leader in the spot market and contribute to decrease the spot price; this is considered as a Pareto-improving change. On the other hand, when the storage cost increases, it is better for the regulator avoiding this "overstoring" practice. Whenever spot markets are not very liquid (that is the case in the European Union), an optimal strategy for the regulator may consist in allowing operators to strategically manage the storage. This kind of strategy is a mean for increasing the quantities of gas available in the markets in order to reduce the price of gas resources.

3.5 Vertical Integration of Storage Facilities

As already noted, in the intermediate market, the activity of storage influences the deregulation process in energy markets. Consequently, the European Commission is implementing Third Party Access to natural gas storage facilities in order to stimulate competition and to promote entry into deregulated markets by new actors. Moreover, recent regulatory reforms in the gas sectors establish legal and functional unbundling of storage system operators who are part of supply undertaking. The main idea is that fair and non discriminatory Third Party Access cannot be achieved by vertically integrated gas operators (mainly historical incumbents) because they would have too high incentives to strategically manipulate storage conditions (tariffs, volumes ...).

In our basic benchmark model, unbundling of storage system applies. Indeed, the storage operator is legally and functionally separated from any downstream supplier, so we can refer to Sect. 3.3 as the unbundling situation. Intuitively, it is commonly expected that an integrated firm may act as a dominant firm in the industry because of the competitive advantage created. What is the effect of vertical integration of storage facilities on strategic storage behaviors of operators? To see this, we consider that storage facilities are integrated to firm $i = 1$ so adding (3.2) and (3.3), its profit is now given by

$$\Pi_1' = p(Q)q_1 - (c+\gamma)y_1 - sz_1 + (a-c)y_2. \tag{3.9}$$

As a result, taking into account this vertical integration assumption implies solving an asymmetric version of the first stage of the basic game. Doing so, we can state the following result.

Proposition 3. *Vertical integration of storage facilities may lead to strategic storage for the integrated operator if the access price is high. However, strategic storage is prevented when the access price is regulated to the socially optimal level.*

Proposition 3 shows that vertical integration of storage facilities can be a source of market power if the access price is quite high. Indeed because of access revenues he earns from the storage activities, the incumbent is driven to increase quantities it stores as the access price increases. Furthermore when the access price is set far enough from the cost, the competitive advantage is so important that the integrated firm is able to sell some volumes on the spot market making the spot price decreasing. As a result, one could argue (as for the case of preemptive access in the previous Section) that strategic storage can be beneficial from a social point of view. Indeed setting a high level of access price, the regulator may have to trade-off more gas availability in the short-term (i.e. in the spot market) against competitive dominance from the (partially) integrated gas firm. It turns out that it is not socially optimal to allow for strategic storage hence low levels of the regulated access price are to be set. This comes from the fact that reducing double marginalization at the storage level is a part of the optimal policy for the regulator. Here vertical cost economies overwhelm gains from gas availability.

However, Breton and Kharbach (2008) have shown that if vertical integration is correlated with preemptive access to storage facilities then, it is also socially preferred to the unbundling situation. This result can be related to our Proposition 2, where strategic storage is not conflicting with the optimal access policy. In a close related model, Baranes et al. (2005) adopt the framework of "strategic purchases" developed by Gaudet and Van Long (1996), in order to focus on the strategic aspects of storage in gas sector. This model has focused on situations where access to storage facilities allows rival firms to adjust strategically the gas price on downstream market. Such a situation occurs when the competitive suppliers are integrated with an upstream oil and gas company. It has been shown that Third Party Access to storage facilities allows the vertically integrated firm (active in both upstream and downstream markets) to strategically rise the intermediate market price in order to increase the cost of the downstream independent firm. In these cases, it seems better from the point of view of welfare to alleviate this strategical behavior by means of the vertical integration of storage operator (when access to storage is opened).

Let us recast these results in our present model assuming that gas firm $i = 2$ is vertically integrated. Now the vertically integrated gas firm $i = 2$ is supposed selling gas through long-term contracts at a market price γ. This assumption leads to reconsider the exogeneity of the supply cost γ. Due to our duopolistic structure, the price is then set in a monopolistic manner in a prior stage. The profit function of firm $i = 2$ then rewrites

$$\Pi_2'(q_2, q_2, y_2, y_1) = p(Q)q_2 - sz_2 - ay_2 + \gamma(y_1 + w_1 + w_2), \tag{3.10}$$

in which we can see in the last part upstream revenues from sellings[15] to traders and the independent firm $i = 1$. Solving for the entire game which has now four stages, the following result can be stated.

Proposition 4. *An integrated gas firm (i.e. $i = 2$) raises its rivals' cost using strategic storage whenever the access price is sufficiently low. Vertical integration of storage facilities to the independent firm (i.e. $i = 1$) may be a counterpart to alleviate such a strategic storage from the upstream integrated gas firm.*

To some extent, patterns of storage strategies and supplies are quite similar to those of Fig. 3.2 which were depicting a leadership configuration. Storage facilities are strategically used by the oil and gas companies active in the upstream and downstream markets. In this context, access to storage facilities allows integrated firms to adjust strategically gas prices on downstream and upstream markets. It allows raising the cost of rival firms that buy natural gas in order to supply the downstream market. This is also a raising rival's cost strategy that induces some distortions and reduces welfare. Integrating storage facilities to a downstream supplier allows to reduce this distortion and to improve the welfare.[16]

Considering these strategic behaviors of oil and gas companies with respect to storage activities, it could be socially optimal for the regulator not to proceed with the unbundling of such activities. Such a partial integration allows to compensate for strategic behaviors of gas firms with upstream market power. More precisely, it lowers the cost supported by the non integrated supplier with regard to that of the vertically integrated oil and gas companies. This idea translates the principle of a symmetric regulation: if authorities accept vertical integration of oil and gas companies, they may also accept vertical integration of the storage activity.

3.6 Flexibility Degree of Storage Facilities

In this Section, we consider more precisely storage facilities which can be used by operators. We assume the same general framework than the basic model presented in Sect. 3.3 introducing two alternative storage facilities and we analyze how it affects behaviors of storage operators.

More precisely, we assume that both facilities differ from operating flexibility related to injection and withdrawal scheduling for gas storage facilities. Without loss of generality, we consider a long-term and the short-term storage facilities. Gas stored in the long-term facility can be only exploited to serve final market whereas operators can use the quantity of gas stored in short-term facility to operate in spot market too.[17] Hence, operator i can inject x_i units of gas in the long-term facility and

[15] The gas production cost is normalized to zero.

[16] A formal proof of this argument is in Baranes et al. (2005).

[17] Underground storages have different technical constraints (storage capacities, working gas volumes, flow rates,...) that depend on physical characteristics. Aquifers storages or depleted fields

y_i units in the short-term facility and, bears a no discriminatory unit cost for access to storage facilities a.

Profit function for a downstream firm i is now given by:

$$\Pi_i(q_i, q_{-i}, x_i, x_{-i}, y_i, y_{-i}) = p(Q+X)(x_i+q_i) - a(x_i+y_i) - sz_i \tag{3.11}$$

where $X = x_1 + x_2$ corresponds to the quantity of gas stored in the long-term facility by both downstream firms.

Here, it should be noted that demand on downstream market is determined from gas quantities stored in long-term facilities and gas quantities withdrawn from the short-term facility.

As in Sect. 3.3, we look at subgame perfect equilibria of the three-stage game. Solving gas retailers competition subgame (choosing q_i) leads to distinguish different cases according to equilibrium features (i.e. interior or corner solutions). Interior equilibrium corresponds to a situation in which both downstream firms offer gas from the short-term facility ($q_i > 0$). On the contrary, a corner equilibrium describes a situation in which a least one firm offers gas exclusively from the long-term facility and without using any short-term facility ($q_i = 0$ for $i = 1$ or $i = 2$ or $i = 1,2$).

The following proposition shows how different levels of flexibility degree for storage facilities affect operators' strategies.

Proposition 5. *At equilibrium, downstream firms choose strategically their offers and storage decisions according to the level of access cost for storage facilities:*

(1) if $a > \frac{11}{27}$, a "Cournot-spot" symmetric equilibrium arises in which downstream firms offer gas exclusively from spot market without using storage facilities
(2) if $\frac{5}{21} \leq a \leq \frac{11}{27}$, two equilibria arise:

(i) the benchmark equilibrium as depicted in Proposition 1,
(ii) a "Stackelberg-like" equilibrium where one downstream firm acts as a leader using only the long-term facility and the rival is excluded from storage and constrained to offer gas from spot market

(3) if $a < \frac{5}{21}$, the benchmark and a "Stackelberg-like" equilibria arise. For the latter, the leader offer gas using the long-term facility and the follower uses both storage facilities and secures its sourcing buying gas on spot market.

Proposition 5 shows how the level of access price to storage facilities affects firms decisions on downstream market and gas sourcing. The most interesting result is that asymmetric equilibria may arise if access price is relatively low ($a \leq \frac{11}{27}$). More precisely, "Stackelberg-like" equilibria may arise with different storage strategies depending on the access price (i.e. higher or lower than $\frac{5}{21}$). In fact, results (2) and (3) show that Stackelberg outcomes can emerge when symmetric downstream firms compete on gas market with both long-term and short-term storage facilities. In each

are used during all the winter period (withdrawal period) to cover base loads; on the contrary, the salt caverns allow instantaneous withdrawals and can be used therefore to serve peak demand.

outcome, one firm behaves as a leader by sourcing only from long-term facilities while the other might be excluded from storage facilities (if $a > \frac{5}{21}$) or pushed to secure its sourcing buying gas on spot market which is relatively more expensive (if $a < \frac{5}{21}$).

Here, Stackelberg equilibria imply that some strategic use of storage occurs: storage helps gas firms to create endogenous leadership using little flexible facilities. As a result, rivals may be pushed to raise their sourcing cost using both spot market and more flexible storage facilities. These results are in the spirit of Saloner (1987) and Pal (1991) who consider a symmetric duopoly with two production periods and show that asymmetric equilibria may arise. Unlike these framework, our results show that Stackelberg equilibria emerge from the imperfect substitutability between gas storage facilities. That is to say, coexistence of long-term and short-term storage facilities gives opportunity for firms to gain market power on downstream gas market.

3.7 Gas Storage in France

The French natural gas market is organized with a high dependency from outside: gas volumes produced in the Southwest (Lacq) are very low and represent for 2007 only 2.2% of internal consumption (CRE, 2008). According to the statistics published by the General Directorate for Energy and Raw Materials in France (DGEMP, 2007), volumes of gas imported with long-term contracts represent 87% of the gas entering in France and come mainly from Norway (31.9%), from the Netherlands (18.8%), from Algeria (18.1%) and from Russia (13.8%); some volumes of gas are also contracted with Egypt (2.7%); Nigeria (1.1%) and Qatar (0.7%). The two incumbent Operators GdFSuez and Total hold almost the entire capacity of gas entering France: in 2007, 89.3% of the gas imported into France was imported by GdFSuez whereas 6% was imported by Total (CRE, 2008). In this context of highly degree of gas dependency from outside, according to Fig. 3.3, France has a very high level of storage to fulfil a high degree of security of gas supply.[18]

More precisely, the very low level of inland natural-gas production explains a high degree of gas dependence from imports and an ample level of cyclical storage.

In this Section, we present the characteristics of the Gas Underground Storage facilities in France (Sect. 3.7.1). The regulatory framework for the Third Party Access to storage facilities in France is described (Sect. 3.7.2) and we focus in Sect. 3.7.3 on the specific role of storage in a context of a high level of gas-supply security. In Sect. 3.7.4, we present the framework and the principles applied to allocate storage capacities for customers. Sect. 3.7.5 explains the structure of tariffs fixed by the owners of underground storages in the form of a negotiated access system.

[18] Data on consumption prior to 2005 are not provided by the Eurostat database.

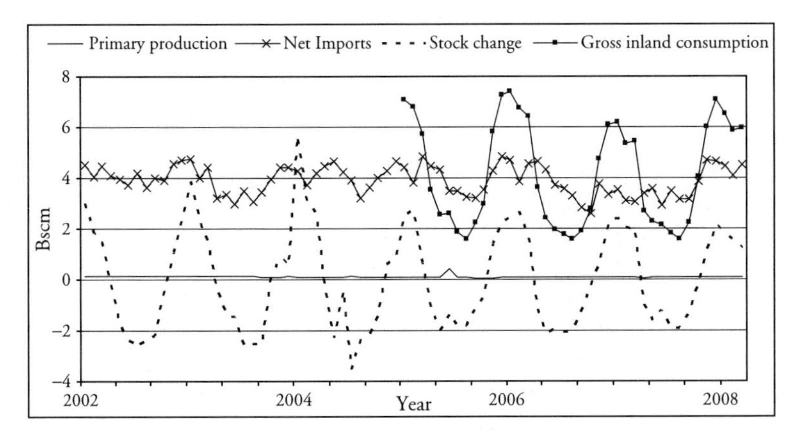

Fig. 3.3 Production, consumption and storage of natural gas in France: 2002–2008. Source: Eurostat

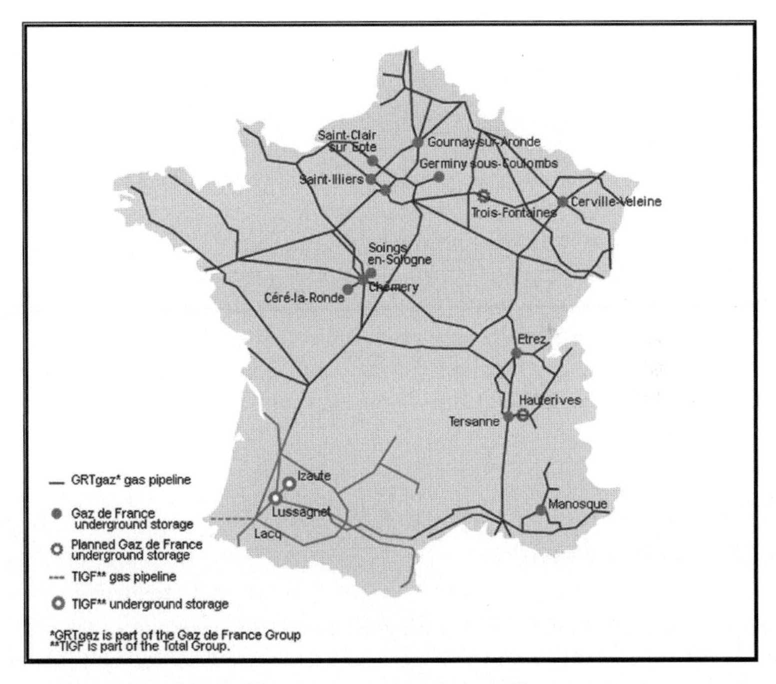

Fig. 3.4 Sites for gas storage in France (source: http://www.gdfsuez.com)

3.7.1 Natural Gas Underground Storage Facilities in France

The French natural gas market is characterized by a high level of storage capacity in relation to other European countries: the capacity of gas storage represents nearly 25% of the final consumption of gas. More precisely, 14 sites of gas storage are localized in France as represented on the map in Fig. 3.4.

Table 3.1 Parameters of storage groups for the storage year 2008–2009

Storage Groups	Number of days of Withdrawal Service	Number of days of Injection Service
Centre	82	110
Ile-de-France Nord t	104	115
Ile-de-France Sud	42	62
Lorraine	77	80
Salins Sud (Salt Caverns)	108	105
Picardie	47	137

Source: http://www.grandesinfrastructures.gdfsuez.com The storage year 2008-2009 runs from 1 April 2008 to 31 March 2009

In France, there are two underground storage system operators that are vertically separated[19]: GdFSuez DGI[20] (the Major Infrastructures Division of GdFSuez) has 12 storage sites which represent 80% of gas storage capacity in France and TIGF (a total subsidiary managing for gas infrastructures) has two storage sites in the Southwest (Lussagnet and Izaute). These underground storage facilities are essentially used for seasonal balancing. (11 sites); only three sites in the Southeast (Manosque, Tersanne and Etrez on the map) are Salt Caverns that allow more flexibility concerning injection and withdrawal services.

These Underground Storage Facilities are organized in Storage Groups. Each storage group is defined through technical parameters such as the number of days for withdrawal service or the number of days for injection service. We give in Table 3.1 the characteristics of the Storage Groups owned and managed by GdFSuez DGI.

On 1 April 2008, 22 gas suppliers subscribed to storage capacity with GdFSuez DGI and 8 gas suppliers subscribed to storage capacity with TIGF.

3.7.2 *The Regulatory Framework for the Third Party Access to Storage Facilities*

As noted in Esnault (2003), "in importing countries, the upstream of the gas chain is not flexible while the demand of gas is characterized by seasonal, daily and hourly variations". In that context, underground storage facilities have several roles: managing seasonality (seasonal storage requirements), managing peaks, security of supply in case of interrupting, optimization of transport infrastructures, etc. These

[19] It is important to note that there is just a legal unbundling in France and no ownership unbundling.

[20] GdFSuez Direction des Grandes Infrastructures.

uses of underground storages are precisely and legally defined and are managed by the incumbent actors GdFSuez DGI and TIGF. As we will see in Sect. 3.7.4, the excess available capacities of storage can be marketed. In that case, these volumes of gas stored can be used to serve the demand or can be withdrawn to be sold on the spot market with an objective to make profits arising from gas price variations between the period of injection service and the period of withdrawal service.

The Annual Report of CRE (2008) presents precisely the regulatory framework for the Third Party Access to underground storage facilities in France. The European Directive of 26 June 2003 leaves the choice between regulated access and negotiated access for underground storage facilities to Member States. France opted for negotiated access under the Law of 9 August 2004. The Ministerial Decree of 21 August 2006 states the general framework of Access to underground storage facilities. The Ministerial Order of 8 February 2008 states precisely storage rights for gas suppliers in relation with their profiles for the storage year 2008–2009.

According to the above mentioned Decree and Order, Major Infrastructures Division of GdFSuez and TIGF organize the access to their underground storage facilities through a "Storage Capacity Allocation Rules" published on their web sites. These two official documents have to be ratified by the French Ministry of Energy.[21]

3.7.3 The Specific Role of Storage for Security of Gas Supply

Gas storage allows security of gas supply in case of interrupting. In that context, storage facilities allow to constitute strategic reserve volumes in order to allow a high level of security. The European Union has precisely defined the principles of security of gas supply.

The Green Paper on security of energy supply[22] focuses on the high level of dependence on gas imports from sources outside the European Union. In this respect, the Gas Directive (2003/55/EC)[23] recognizes the right of Member States to manage security of supply as a Public Service Obligation. In this context, the Ministerial Decree of 19 March 2004 (Decree N° 2004-251) defines precisely the public service obligations in the French gas sector. The Article 4 of the mentioned Decree dictates that France should run an extremely high degree of security of gas-supply. More precisely, the suppliers have the obligation to supply natural gas as a continuous

---

[21] These two official documents can be downloaded on the web sites of the Division of Infrastructures of GdFSuez and Total (http://www.grandesinfrastructures.gdfsuez.com and http://www.tigf.fr). For the storage year 2008–2009, these "Storage Capacity Allocation Rules" were published January 23*rd* 2008 for GdFSuez DGI and April 1*st* 2008 for TIGF.

[22] "Towards an European strategy for the security of energy supply", Green Paper, COM/2000/0769 final.

[23] Directive 2003/55/EC of the European Parliament and of the Council of 26 June 2003 concerning Common Rules for the Internal Market in Natural Gas and repealing Directive 98/30/EC.

service even in three extreme events (Article 4 of the Decree) that are: (1) "loss by the supplier during a maximum period of 6 months of its main gas procurement source in the context of average weather conditions", (2) "exceptionally cold winter such as one statistically that occurs every 50 years" and (3) "extremely low temperature during a maximum period of three days, such as one statistically that occurs every 50 years".

In order to fulfil their obligation of uninterrupted supply, suppliers must ensure that they may resort to alternative means as underground storage of gas (Article 5 of the Decree). In that context, a calculation is made in order to define the level of strategic reserve volumes of gas that have to be stored in France.[24]

On 26 April 2004, a Council Directive (2004/67/EC) establishes a common framework within which Member States can define general security-of-supply policies that are transparent and non-discriminatory.[25]

3.7.4 The Principles for the Allocation of Underground Storage Capacities

In accordance with Sect. 3.7.3, the allocation of underground storage capacities is made in a sequential way with a priority system described in the "capacity allocation rules" we sum up trough three points.

First, the Article 3 of the Ministerial Decree N°2006-1034 (August 2006) stipulates that the underground storage facilities are allocated with a priority access to the Transport and Storage Systems' Operators (GdFSuez and TIGF). They have a preemptive access to storage facilities. Specific contracts give them access to the storage capacities needed in order to fulfil the optimization of transport and storage infrastructures in balancing zones.

Second, the remaining storage capacities are allocated to suppliers with an effective final customers' portfolio on the basis of the storage profiles and the unit rights defined in the Ministerial Decree of February 2008. More precisely, all suppliers operating in the French market must store natural gas before the winter in proportions set out by the profile of their customers' portfolio. This allocation of underground storage capacities represents (for the storage year 2008–2009) a reserve volume of 120.66 TWh and a potential daily withdrawal rate of 2,480 GWh per day. On 1 April

[24] As noted in the reports DRI-WEFA (2001a, 2001b and 2002), this calculation is made assuming that "the probability of 1/50 winter together a political supply crisis is so low that it is not reasonable to cover the simultaneous possibility."

[25] The Article 4 of the Directive contains the same principles that are listed in the Ministerial Decree of 19 March 2004 (Decree N° 2004-251). This Article 4 states precisely that Member States will ensure that supplies for household customers are protected at least in the event of:
"a partial disruption of national gas supplies during a period to be determined by Member States taking into account national circumstances; extremely cold temperatures during a nationally determined peak period; periods of exceptionally high gas demand during the coldest weather periods statistically occurring every 20 years".

2008, 101.7 TWh of capacities have been allocated (CRE, 2008). This allocation is used to fulfil Public Service Obligations concerning security of gas supply and allows therefore continuity of service for customers.

Finally, Article 14 of the Ministerial Decree N°2006-1034 states that excess storage capacity is made available to the market under transparent and non discriminatory conditions. In this framework, additional storage capacities have been made available and have been commercialized by GdFSuez DGI. Two sales with a bidding process have been organized in March 6th and 13th, 2008. An other sale was organized on April 10th, 2008 according to the principle "first arrived, first served" rule to allocate additional storage capacities.[26] No sale was organized by TIGF due to the unavailability of excess storage capacities.

3.7.5 *Tariffs for the Use of Storage Facilities*

In accordance with a negotiated access for underground facilities, CRE has no authority over the tariffs fixed by GdFSuez DGI and TIGF. The terms of these access charges are based on storage use and reflect the constraints inherent in storage groups. In the case of a basic storage service, GdFSuez DGI and TIGF propose an access charge with a specific structure.

Under physical constraints for each storage facility (level of working gas volume, constraints on the flows of gas injected or withdrawn, . . .), GdFSuez DGI proposes a two-part tariff (two components). The first part of the tariff includes an amount to be paid by the customer for the capacity reservation according to the characteristics of the storage group. The level of the storage capacity rate is given in the following Table 3.2 for the storage year 2008–2009.

Table 3.2 Nominal storage capacity rates

Storage group	Nominal storage capacity Charge rate (€ per yr per MWh)
Centre	5.50
Ile-de-France Nord	4.80
Ile-de-France Sud	7.90
Lorraine	5.30
Salins Sud	15.25
Picardie	7.10

[26] Almost 6 TWh has been sold during the sale of March 6th, 1 TWh has been sold during the sale of March 13th and the overall 59 GWh of gas storage capacities has been sold during the sale of April 10th. All the details of these sales (especially the prices for Capacity Reservation) are available on the web site of GdFSuez DGI.

The second part of the tariff is a commodity component to be paid relating to the gas volumes injected or withdrawn. The withdrawal charge rate and the injection charge rate are the same for all storage groups owned by GdFSuez DGI: 0.10 € per MWh for the withdrawal charge rate and 0.31 € per MWh for the injection charge rate.

TIGF defines Units of Storage as "fixed bundles" with three components depending on the standard service supplied.[27] In the framework called "dynamic standard service", the Unit of Storage is composed of a Unit of Storage Capacity (USC) (10,000 kWh), a Unit of Daily Withdrawal Capacity (152 kWh per day) and a Unit of Daily Injection Capacity (85 kWh per day). In the framework called "equilibrium standard service", the Unit of Storage is composed of a Unit of Storage Capacity (10,000 kWh), a Unit of Daily Withdrawal Capacity (88 kWh per day) and a Unit of Daily Injection Capacity (84 kWh per day).

The customer subscribes a number of Units to storage capacities with TIGF. The access charge paid by customer is equal to the sum of the three following components. The first component is a Fixed Annual Rate (FAR) equal to 5,000 € per year and independent of the number of Storage Units subscribed; the second is an annual Subscription Rate that is proportional to the number of the Units of Storage subscribed, using the following Storage Unit Prices (Table 3.3). Finally the last component is a charge paid for injection or withdrawal services (depending on the gas injected or withdrawn) with a price of 0.26 € per MWh for withdrawn volumes and a price of 0.16 € per MWh for injected gas quantities.

To conclude the presentation of this French case, we can underline that gas underground storage facilities are mainly used to fulfil Public Service Obligations and specially to fulfil security of gas-supply. In this situation, the General Directory of Competition of the European Union considers that the French storage facilities are not sufficiently opened and that the excess storage available for the customers is too low. As noted by Esnault (2003), "Regarding the rigidity of gas chain, new comers need to use storage facilities to sell gas physically and to trade on new spot markets". More storage capacities could be opened to improve the liquidity of the gas market and to increase the volume of "free gas" in this market (see Baranes et al., 2007). It would be profitable for marketers and could induce a decrease of gas prices in the short run. A development of opening access to storage would induce necessarily lower gas volumes stored for the security of gas supply. In other words, a larger opening to gas storage facilities would allow higher liquidity on the gas market to the detriment of fulfilling the Public Service Obligations. It represents a high

Table 3.3 Storage unit price

Dynamic service	55.65 €
Equilibrium service	30.90 €

[27] Two standard services are supplied by TIGF (Dynamic Service and Equilibrium Service) depending on physical characteristics (the maximum and minimum storage levels, the injection and withdrawal development factors, ...).

political risk in a context of a high degree of gas dependency from outside. In this context, "Storage can assume a more important role for supply security than it did in the traditional world of vertically integrated companies" (von Hirschhausen, 2008).

3.8 Conclusion

We developed a basic model of competition between gas firms (suppliers, traders) which have access to storage facilities. Mainly, this framework helps us to analyze conditions in which storage facilities allow firms exerting market power on downstream gas markets. When there is no discrimination for the access to storage facilities, as it is the case with unbundling, firms do not use storage strategically to get more market power. However when there is a priority access to the storage system, which creates leadership for given operators, strategic storage might occur without reducing social welfare. Similar results appear when storage facilities are integrated to one downstream operator or when some gas firms are vertically integrated along the gas chain. Finally, when long-term and short-term storage facilities coexist firms may gain market power on downstream gas market using storage strategically.

3.9 Appendix

3.9.1 Proof of Proposition 1

Solving the *third stage* of our basic game in order to determine a traditional Cournot symmetric (interior) equilibrium implies maximizing profits, given in the text by (3.2), with respect to q_i. First-order conditions derived from this maximization write for firm i : $\frac{\partial \Pi_i}{\partial q_i} = -2q_i - q_{-i} + 1 - s = 0$ and solving it for $q_1 = q_2$ leads to $q^*(s)$ in the text. As a result at the *second stage* and according to the spot market clearing condition (3.5) in the text we can easily find that $q^{*-1}(x) = 1 - 3x$. The Nash symmetric equilibrium for traders in the spot market is a couple (w_1, w_2) such that at this subgame profits π_j are maximized using Nash conjectures. These profits write $\pi_j(w_j, w_{-j}) = (S\left(\frac{y_1+y_2+w_j+w_{-j}}{2}\right) - \gamma)w_j$ so the first-order condition for trader j is given by $\frac{\partial \pi_j}{\partial w_j} = 1 - 3w_j - \frac{3\left(y_1+y_2+w_{-j}\right)}{2} = 0$. The symmetric equilibrium at this stage is then $w^*(y_1 + y_2) = \frac{2}{9} - \frac{1}{3}(y_1 + y_2)$. Finally, moving back to the *first stage*, for each gas firm $i = 1, 2$, profits are given by $\Pi_i = [p(2q^*(s^*(y_1 + y_2))) - s^*(y_1 + y_2))]\, q^*(s^*(y_1 + y_2)) - (a - s^*(y_1 + y_2))\, y_i$. At the Nash interior equilibrium in storage strategies, first-order conditions write $\frac{\partial \Pi_i}{\partial y_i} = \frac{11}{27} - \frac{17}{18} y_i - \frac{4}{9} y_{-i} - a = 0$ for $i = 1, 2$. Solving them together leads to $y^* = \frac{2}{75}(11 - 27a)$. ■

3.9.2 Proof of Proposition 2

Solving for the storage subgame equilibrium yields a best-reply for the follower given by $\hat{y}_2^R(y_1) = \frac{1}{17}\left(\frac{22}{3} - 8y_1 - 18a\right)$ and equilibrium storage levels for the leader and the follower respectively $\hat{y}_1 = \frac{1}{297}(98 - 290a)$ and $\hat{y}_2 = \hat{y}_2^R(\hat{y}_1) = \frac{1}{297}(82 - 178a)$. The resulting spot price and downstream supplies are then $\hat{s} = \frac{1}{33}(1 + 26a)$ and from $\hat{q} = q^*(\hat{s}) = \frac{2}{99}(16 - 13a)$. Hence one can easily see that $\hat{z}_1 = \hat{q} - \hat{y}_1 = \frac{2}{297}(-1 + 106a) \leq 0$ if $a \leq \frac{1}{106}$ but $\hat{q} - \hat{y}_2 = \frac{2}{297}(7 + 50a) > 0$.

In order to find the regulated level of the access charge to storage facilities, we solve the related problem $\max_{a \geq c} \hat{W}(a,c)$. where $\hat{W}(a,c)$ is the social welfare defined as $\hat{W}(a,c) = \int_0^{2\hat{q}} p(Q)\,dQ - 2\hat{q}p(2\hat{q}) + \Pi_1(\hat{q},\hat{q},\hat{y}_1,\hat{y}_2) + \Pi_2(\hat{q},\hat{q},\hat{y}_2,\hat{y}_1) + 2\pi_1(\hat{w},\hat{w}) + \pi_s$. Here traders equilibrium offers are given by $\hat{w} = \frac{2+52a}{99}$. Denote $\hat{a}$ the optimal access charge with leadership, we find $\hat{a} = \max\{c, \frac{297}{52}c - \frac{35}{52}\}$, where $\hat{a} > c$ if $c > \frac{1}{7}$. So if $c < \frac{1}{106}$, then $\hat{a} = c$ and $\hat{z}_1 < 0$. ∎

3.9.3 Proof of Proposition 3

Again stages 2 and 3 of the game in the basic model remain unchanged and we look for a Nash equilibrium in storage strategies of the first subgame where payoffs are given in the text by (3.9) and (3.2) for firm $i = 2$ with $z_i = q_i + y_i$, $q_i = q^*(s^*(y_1 + y_2))$ and $s = s^*(y_1 + y_2)$. Hence the interior Nash equilibrium is such that $\tilde{y}_1 = \frac{2}{9} + \frac{2}{325}(21a - 96c)$ and $\tilde{y}_2 = \frac{2}{9} + \frac{2}{325}(21c - 96a)$. Then spot market equilibrium price is $\tilde{s} = s^*(\tilde{y}_1 + \tilde{y}_2) = \frac{1}{9} + \frac{3(a+c)}{13}$ so downstream supplies writes $\tilde{q}_i = q^*(\tilde{s}) = \frac{8}{27} - \frac{1}{13}(a + c)$. As a result with $a \geq c$, $\tilde{z}_2 > 0$ and $\tilde{z}_1 < 0$ if $a > \tilde{a}_0 = \frac{650}{1809} + \frac{167c}{67}$. Finally, solving the problem $\max_{a \geq c} \tilde{W}(a,c)$ where $\tilde{W}(a,c)$ is the social welfare with vertical integration leads to $\tilde{a} = \max\{c, \tilde{a}_1\}$ where $\tilde{a}_1 = \frac{31991c}{18241} - \frac{32500}{492507}$. Then one can check that $\tilde{a}_0 > \tilde{a}_1$ so it never occurs that $\tilde{z}_1 < 0$. ∎

3.9.4 Proof of Proposition 4

(a) First assume that there is unbundling of storage facilities (as in the benchmark case). Again compared to our basic model (see 3.9.1), the last stage equilibrium remains unchanged. Notice that the payoff of firm $i = 1$ is given by (3.2). At the third stage, the subgame equilibrium trader positions are now dependent on the upstream gas price γ so that $w_i = \frac{2}{9} - \frac{1}{3}(y_1 + y_2) + \frac{2}{3}\gamma$. Compared to (3.7), the spot market equilibrium price is now incorporating γ as a cost such that $s^*(y_1 + y_2) = \frac{1}{3} - \frac{1}{2}(y_1 + y_2) + \frac{2}{3}\gamma$. Hence at the second stage, the Nash equilibrium in storage strategies can be found to be $y_1^* = \frac{2}{75}(11 - 27a - 19\gamma)$ and $y_2^* = \frac{2}{75}(11 - 27a + 6\gamma)$. At this point to complete the solution, we go up to the first new stage, where firm

$i = 2$ sets the gas price γ^* in such a way that $\gamma^* = \arg\max_\gamma \Pi_2'$ where Π_2' is given by (3.10) in the text. Simple optimization routine gives $\gamma^* = \frac{394}{951} - \frac{86}{317}a$. Finally equilibrium downstream supplies equal $q_i^* = \frac{194}{951} - \frac{52}{317}a$ then $z_1^* > 0$ for all a and $z_2^* = -\frac{148}{951} + \frac{190}{317}a \leq 0$ if $a \leq \frac{74}{285}$. This complete the proof of the first part of the Proposition.

(b) Now just assume that storage facilities are integrated to firm $i = 1$ so that its payoff is then given by (3.9). Following same steps as in point (a) of this Proof, one can find at the end that $\tilde{y}_1^* = \frac{2}{75}(11 + 24a - 19\gamma - 51c)$, $\tilde{y}_2^* = \frac{2}{75}(11 + 24c + 6\gamma - 51a)$, $\tilde{\gamma}^* = \frac{3}{1901}(223 + 182a - 493c)$. Hence $\tilde{z}_1^* > 0$ for all $a < \frac{26807}{198416} + \frac{145503}{99208}c$ and $\tilde{z}_2^* \leq 0$ if $a \leq \frac{18275}{107448} + \frac{2741}{35846}c$.

(c) To complete the proof, just see that $\tilde{z}_2^* - z_2^*$ is a strictly increasing linear function of a and strictly positive when $a = c$ which proves that $\tilde{z}_2^* > z_2^*$ for all $a \geq c$.
■

3.9.5 *Proof of Proposition 5*

(i) Solving the third stage in order to determine the Cournot equilibrium implies maximizing profits given in the text by (3.11) with respect to q_i. According to parameter values both interior ($q_i > 0$ for all $i = 1, 2$) and corner equilibrium ($q_i = 0$ and $q_j > 0$) may arise. In both cases, we express parameter values for which a given equilibrium arises. Best replies for an interior equilibrium are given by $q_1(q_2) = \frac{1}{2}(1 - s - x_2 - q_2) - x_1$ and $q_2(q_1) = \frac{1}{2}(1 - s - x_1 - q_1) - x_2$ and solving it leads to $q_i^*(s, X) = \frac{1}{3}(1 - x_{-i} - 2x_i - s)$. As a result at the second stage now $S(\xi) = 1 - 3\xi$ with $\xi = X + Y + W$, and we have $w^*(X, Y) = \frac{2}{9} - \frac{1}{3}(X + Y)$ with $Y = y_1 + y_2$. Finally, moving back to the first stage, for each gas firm $i = 1, 2$, we obtain Nash equilibrium in storage strategies: y^*(as given in benchmark) and $x_1^* = x_2^* = 0$. This equilibrium arises when $a \leq \frac{11}{27}$.
(ii and iii). For corner solutions, let consider the case $q_1 = 0$ and $q_2 > 0$. Solving the third stage and looking for a corner solutions give $q_2^*(s, X) = \frac{1}{2}(1 - x_1 - s) - x_2$. At the second stage and according to the spot market clearing (where now $S(\xi) = 1 - 2\xi$ with $\xi = \frac{x_1}{2} + x_2 + Y + W$), Nash equilibrium for traders is such that $w^*(X, Y) = \frac{1}{6}(1 - x_1 - 2x_2) - \frac{1}{3}Y$. Moving back to the first stage, we have $\{y_1^* = 0, x_1^* = \frac{1}{2} - \frac{3a}{4}, x_2^* + y_2^* = 0\}$ or $\{y_1^* = 0, x_1^* = \frac{3}{7} - \frac{3a}{5}, x_2^* + y_2^* = \frac{2}{7} - \frac{3a}{5}\}$, according to wether $a \lesseqgtr \frac{5}{21}$. Finally, note that similar symmetric equilibria exist with $q_2 = 0$ and $q_1 > 0$. ■

References

Arvan, L. (1985). Some examples of dynamic cournot duopoly with inventories. *The Rand Journal of Economics, 16*, 569–578.

Baranes, E., Mirabel, F., and Poudou, J. C. (2005). Storage and competition in gas market. *Economics Bulletin, 12*(19), 1–9.

Baranes, E., Mirabel, F., and Poudou, J. C. (2007). Strategic storage and competition in European gas markets. *Cahier de Recherche du Laser*, n° 24-09-08.

Breton M. and Kharbach, M. (2008). The welfare effects of unbundling gas storage and distribution. *Energy Economics, 30*, 732–747.

CRE, (2008). *Activity Report 2008*, French Regulation Commission for Energy, July 2008.

DGEMP, (2007). *France's Energy Situation for 2007*, General Directorate for Energy and Raw Materials.

DRI-WEFA, (2001a). Report for the European Commission: results from opening the gas market, Volume I, *European Overview*, report made for Directorate General for Transport and Energy, July 2001.

DRI-WEFA, (2001b). Report for the European commission: Results from opening the gas market, Volume II, *Country reports*, report made for Directorate General for Transport and Energy, July 2001.

DRI-WEFA, (2002). European gas storage at a crossroads forecasts through 2020, *Report*, September 2002.

Durand-Viel, L. (2007). Strategic storage and market power in the natural gas market. *mimeo*.

Cremer H. and Laffont J. -J., (2002). Competition in gas markets. *European Economic Review, 46*(4–5), 928–935.

Egging, R. G. and Gabriel, S. A., (2006). Examining market power in the European natural gas market. *Energy Policy, 34*, 2762–2778.

Esnault, B. (2003). The need for regulation of gas storage: The case of France, *Energy Policy, 31*, 167–174.

Kirman, A. P. and Sobel, M. J. (1974). Dynamic oligopoly with inventories. *Econometrica, 42*, 279–287.

Modjtahedia, B. and Movassagh, N. (2005). Natural-gas futures: Bias, predictive performance, and th,e theory of storage. *Energy Economics, 27*, 617– 637.

Pal, D. (1991). Cournot Duopoly with Two Production Periods and Cost Differentials. *Journal of Economic Theory, 55*, 441–448.

Pal, D. (1996). Endogenous stackelberg equilibria with identical firms. *Games and Economic Behavior, 12*, 81–94.

Philps, L. and Richard J. F. (1989). A dynamic oligopoly model with demand inertia and inventories. *Mathematical Social Sciences, 18*, 225–243.

Poddar S. and Sasaki D. (2002). The strategic benefit from advance production. *European Journal of Political Economy, 18*(3), 579–595.

Saloner, G. (1987). Cournot duopoly with two production periods. *Journal of Economic Theory, 42*, 183–187.

Salop S. G. and Scheffman, D. T. (1983). Raising rival's cost. *American Economic Review, 73*(2), 267–271.

von Hirschhausen, C. (2008). European supply security in natural gas (CESSA Policy Brief), August 17th, Dresden University, EU Energy Policy Blog.

Mu, X. (2007). Weather, storage, and natural gas price dynamics: Fundamentals and volatility. *Energy Economics, 29*, 46–63.

Chapter 4
The Regulation of Access to Gas Storage

Alberto Cavaliere

4.1 Introduction

Due to the implementation of directives 98/30/EC and 2003/55/EC, during the last decade natural gas markets have been liberalised in the European Union. National regulatory reforms have been carried out in order to implement unbundling and non discriminatory third party access to essential facilities. At present each member Country is expected to implement legal unbundling of transmission and distribution networks from potentially competitive activities, though the European Commission is now fostering the introduction a third "liberalisation package" in order to strengthen unbundling requirements. In order to let new entrants compete with former integrated utilities on a level playing field, regulated third party access to transmission and distribution networks has also been imposed by the last EC directive.

A particular feature that differentiates natural gas from electricity is the possibility of storage. Gas consumption is affected by seasonal, weekly and daily fluctuations, both predictable and unpredictable. Utilities need to constantly balance demand and supply (which might be flat). Access to storage gives suppliers the flexibility needed to cope with demand uncertainty. Therefore directive 2003/55/EC also requires unbundling of gas storage and non discriminatory third party access to storage facilities. However, member Countries are just required to implement accountancy unbundling of storage assets and can opt between negotiated and regulated third party access, according to the features of national storage markets.

Though storage costs are affected by scale economies, storage is not a natural monopoly. Any storage plant can supply storage services in competition with other existing plants as minimum efficient scale is generally far from the amount of total storage demand from gas suppliers. However the liberalisation directives did

A. Cavaliere
Università degli Studi di Pavia, Facoltà di Economia, Via San Felice, 5 27100, Pavia, Italy
e-mail: alberto.cavaliere@unipv.it

A. Cretì (ed.), *The Economics of Natural Gas Storage: A European Perspective*,
DOI: 10.1007/978-3-540-79407-3_4,

not require divestiture of existing storage assets owned by former integrated utilities, in order to introduce storage to storage competition. At present a competitive market for storage is effective only in the UK, where ownership unbundling has already been implemented and multiple storage companies operate their business independently from gas supply, under the supervision of antitrust authorities. Most continental Countries are characterized either by de facto monopolies or by market power in the storage sector. Moreover, access to storage facilities is still granted by branches of the former integrated gas utilities,[1] now operating as dominant gas suppliers in the downstream market.

However storage services are not the unique flexibility source for gas suppliers. Flexible production fields and flexible importing contracts may operate as a substitute for gas storage, as well as interruptible contracts with industrial customers or access to spot market for gas (though not yet liquid enough in Continental Europe). But in practice these storage substitutes can hardly meet the demand for flexible gas by any supplier. Moreover the nature of most flexibility inputs is such that a market for flexibility is hard to define. Flexible production fields are available just in gas producing Countries like the UK and the Netherlands. Flexible import contracts and a sufficient portfolio of interruptible contracts with industrial customers are generally positively correlated with market shares and therefore more available to incumbents than to new entrants. Moreover, even if the duplication of storage plants were considered economically viable by new entrants, it would require suitable sites and a long time span to carry out new investments.

Once we consider existing storage plants as essential facilities to compete in the gas market the need to regulate access prices ex-ante follows, as non discriminatory third party access (from now onwards TPA) – implemented through negotiated tariffs – may not be sufficient to control market power. However, the regulatory setting should account for the availability of storage substitutes and their asymmetric distribution among gas suppliers. Standard regulation by cost-reflective storage tariffs reviewed with price-cap mechanisms not necessarily leads to an efficient allocation of the existing storage resources among gas sellers, when the latter differ over the availability of storage substitutes. The need to consider the availability of storage substitutes beyond the degree of concentration in local storage markets has also been recently recognized by the Federal Energy Regulatory Commission (FERC) in the USA (FERC, 2005). Considering also storage substitutes is deemed useful to better assess market power when storage companies ask FERC an exemption from regulated tariffs, claiming they operate in a competitive storage market.

Furthermore the market for storage is at present affected by capacity constraints in most European Countries. According to an inquiry carried out by Energy Regulators, in 2006 43.5% of total European storage capacity was fully booked. As for another 38% of the total capacity, the inquiry found that less than 5% of technical

[1] When liberalisation took place national markets were characterized by just one or a few companies owning multiple storage plants and European directives did not impose any horizontal unbundling aiming to split storage companies by selling part of their plants to new entrants, as was done in the case of electricity generation.

capacity was available (ERGEG, 2006). The top 15 storage operators state that actual capacity is insufficient compared to the demand for storage services. Scarcity is even expected to increase in the future.[2] The most recent inquiry carried out by the European Commission (European Commission, 2007) has also found that access to storage is foreclosed by long-term reservations and capacity hoarding. In fact, due to the absence of "used-it-or-lose-it" provisions, booked storage is not fully used.[3] In some Countries like the Netherlands and Denmark, gas suppliers can overcome bottlenecks by resorting to plants located in Germany.

Due to capacity constraints, the efficiency properties of the allocation of storage resources depends on rationing criteria implemented by regulators or storage companies through congestion management rules. Efficiency could be pursued by allocating scarce storage capacity according to its value for any single gas supplier. We expect these values to be heterogeneous due to the asymmetric availability of storage substitutes. However at present this principle seems not to be respected by access rules implemented in practice. Inefficient allocation rules may in turn produce distortions not only in the storage market but also in the downstream market for gas supplies. The same inquiry quoted above has highlighted that wholesale and retail markets for gas are still affected by the market power of incumbents, which operate as dominant firms after liberalisation. The availability and distribution of storage resources is considered one of the causes of market foreclosure in Europe.

We claim that in order to discuss the efficiency of rationing rules storage should be considered along with all the inputs that provide flexibility to gas suppliers. The asymmetric distribution of storage substitutes between incumbents and new entrants, as well as the long time span required to deliver new capacity to the market lead to recognize storage plants as essential facilities, even if storage is not a natural monopoly and duplication of storage facilities may be viable in principle. The need to regulate storage tariffs then follows. Though charging cost reflective access tariffs may be effective in controlling the exercise of market power by *de facto* storage monopolists, such a choice not necessarily leads to an efficient allocation of storage capacities. Regulated prices may not signal storage scarcity and the final allocation of storage resources depends on the rationing rule arising from congestion management.

The efficiency properties of this rule require the allocation of storage according to its idiosyncratic value for each gas supplier. Therefore we find that an efficient rationing rule should equalize the shadow price of storage across gas suppliers. The shadow price accounts both for the quantity of output to be supplied by each

---

[2] Recent forecasts concerning North–Western Europe (Hoffler and Kubler, 2006) show that a storage gap is going to affect the whole region in ten years, due also to the expected decrease of national production in the UK and the Netherlands. The Storage gap could be even wider than expected if the increasing import dependency led more Countries to devote storage capacity to precautionary inventories.

[3] The inquiry has found that most of the storage from the sample which is fully booked has been more than 95% full at the beginning of winter (in the period from January 2003 to mid-2005). In some cases however less than 90% of capacity has actually been used (European Commission, 2007, p. 65).

company and the cost of its storage substitutes. Current rules adopted in Continental Europe appear then to be inefficient, as they allocate storage either just according to the market shares of gas suppliers in the household market (Italy) or to a "first come, first served" criterion (in Germany, Denmark and The Netherlands for instance). In our contribution we have explored the effects of inefficient rationing rules in the case of the Italian gas market which is structurally affected by storage scarcity and is regarded as a heavily regulated storage market. Though the Italian storage monopolist is considered to be the main storage company across Europe, excess demand for seasonal storage persists since liberalization started. Moreover, though the lack of investment and the huge amount of precautionary stocks contribute to storage scarcity we claim that the rationing rule adopted by the regulator to allocate existing capacity may also be responsible for this undesirable result. Recent empirical research seems to confirm this claim.

However, even ex-ante efficient rules may be difficult to implement due to asymmetric information about the technology of flexibility. To the extent that gas suppliers are rationed they are led to report to the regulator larger storage requirements than those implied by the equalization of shadow prices. Moreover, storage demands should be assessed not only according to their effect on the productive efficiency of gas suppliers, considered in isolation from each other, but also taking account of the structure of the final gas market. If we assume imperfect competition in the final gas market, then storage demands may be affected by strategic behaviour. In a market structure with a leader that controls the final gas price and a competitive follower, we show that the adverse incentives of the dominant firm lead it to demand storage with the aim of excluding the follower from the final market. Then "first come, first served" rules and long-term booking of storage capacity can be considered particularly detrimental to competition, as they allow the dominant firm to carry out exclusionary strategies aimed at market monopolization.

At a first glance market mechanisms like auctions should perform better with respect to efficiency goals, to the extent they are expected to allocate storage according to its value for bidders. We have then compared general (non-specified) administrative rules with a market mechanism that allocates storage capacity according to gas suppliers bids. The market mechanism we have considered is a *multiunit sealed bid uniform price auction* that raises the price of storage above the regulated cost-reflective tariff: storage capacity is allocated by charging the market clearing price arising out of the bidding equilibrium. The dominant firm is again assumed to behave strategically, driving the market clearing price towards the level that optimizes – from its point of view – the quantity of storage assigned to the follower. However, to the extent that the dominant firm is charged the same price to get its portion of storage capacity, the market mechanism makes the exclusionary behavior more costly and potentially increases the portion of capacity left to the follower, as compared with an administrative allocation that just charges the cost-reflective access price. Since this market mechanism welfare dominates any administrative distribution of storage capacity, except when the latter leads the follower to get more storage capacity, welfare analysis leads to prefer the former mechanism. In fact, in our setting any increase of the portion of storage capacity allocated to the follower is pro-competitive and allows to increase both consumer surplus, social welfare and the follower profits, though it reduces the profits of the dominant firm. Obviously the profits of the storage company are increased by the adoption of a market mechanism due to the rents that accrue to it through a market price greater than the cost-reflective regulated tariff.

After reviewing the scarce literature on regulation of gas storage (Sect. 4.2), we will consider to what extent storage plants may be considered as essential facilities, applying a test developed by the antitrust practice (Sect. 4.3). In Sect. 4.4, we will give a theoretical representation of the technology of gas supply, assuming that flexibility may be obtained by an array of inputs including access to storage capacity. In this framework we can analyse the effects of storage rationing on the productive efficiency of gas suppliers and distinguish between the cost and the value of storage, the latter being represented as its shadow price. In Sect. 4.5, we will consider a specification of the gas supply technology and derive an example of optimal rationing rule for the allocation of storage capacity, accounting for the availability of storage substitutes. In Sect. 4.6, through equilibrium analysis, we will consider the effects of the rationing rule on competition in the downstream market and compare the allocation of storage implemented by the regulator through administrative rules with the allocation resulting from a market mechanism (storage auctions). In Sect. 4.7, the same comparison is made on welfare grounds. In Sect. 4.8 we will consider the Italian experience as an example of the effects of inefficient allocations of storage in the gas market. Some conclusions will follow (Sect. 4.9).

4.2 Regulation of Access to Storage: Related Literature

To the best of our knowledge, regulatory issues concerning gas storage have never been tackled by economic theory. However some recent contributions have analysed the economics of gas storage and the effects of liberalisation policies carried out in Europe. Chaton, Cretì and Villeneuve (in press) deal with seasonal gas storage in a competitive gas market, where injection and withdrawal decisions react to price fluctuations, taking account the depletion of exhaustible gas resources. Cretì and Villeneuve, in Chap. 5, deal with precautionary gas storage as an instrument to face supply disruptions, considering market failures with respect to security of supply goals. The strategic nature of storage decisions in an imperfectly competitive gas market is considered by Baranes, Mirabel and Poudou in Chap. 3 and by Durand-Viel (2007). According to Baranes, Mirabel and Poudou, TPA to independent storage facilities can reduce welfare because it incites firms that are vertically integrated upstream to make storage decisions that increase wholesale gas prices for downstream competitors (storage may be used to raise rivals' cost). Therefore fostering integration of storage facilities with downstream suppliers is considered to be a better solution then TPA, from a social welfare the point of view. Both in Baranes, Mirabel and Poudou (Chap. 3 of this book) and in Durand-Viel (2007) the alternative between carrying inventories or buying the commodity in the spot market is analysed. However the two models differ in the assumptions about market structure at the production stage. Assuming oligopoly both at the production and supplier (wholesale) stage, Durand-Viel considers storage not only as an instrument to preempt future demand but also as a strategic tool for suppliers to deter producers' market power. Actually storage decisions can avoid the increase of prices in the spot market but, given that all downstream competitors share this benefit, storage by any firm exerts a positive externality on rival suppliers. Strategic storage decisions should therefore take account of this trade-off.

In our contribution we will focus on regulatory issues by considering both the existence of storage substitutes and congestion management due to capacity constraints. Moreover the efficiency of rationing rules concerning access to storage will be considered not only per se but also with reference to the distortions induced on the downstream market. Most of the above quoted contributions assume the existence of a liquid spot market, which is not a realistic assumption at the present liberalisation stage in Continental Europe, where most gas exchanges take place through long-term contracts with *take or pay* clauses. Therefore we assume that the downstream market is characterized by a dominant firm and a competitive fringe of new entrants behaving collectively as followers vis-à-vis the incumbent. In our opinion such an assumption better fits the result of the recent inquiry by the European Commission about competition in European gas markets (European Commission, 2007).

4.3 Is Gas Storage an *Essential Facility*?

In liberalized gas markets storage is also a tool for price arbitration, be it seasonal (summer–winter price differentials can easily be exploited by resorting to gas storage) or between the electricity and the gas markets (if price soars at the power exchange, utilities can obtain rents from gas fired power plants by resorting to gas in storage instead of buying it in the spot market at a higher price, provided that the cost of access to storage is not too high). Therefore due to multiple and rival uses of gas storage, capacity constraints may easily arise, especially at the start of liberalisation and considering that network safety and security of supply goals are obviously given a higher priority than the commercial exploitation of storage capacity.

However flexibility resources are not restricted to storage capacity. Countries like the UK can rely on indigenous production fields whose output may fluctuate according to demand ("supply swing"), therefore their need for seasonal gas storage is reduced. Suppliers in importing Countries can profit from flexibility clauses of long-term contracts, allowing to maximise imports up to capacity during peak times and to reduce them when demand is weaker. Liquid spot markets provide an additional flexibility source for gas sellers. However these flexibility sources can hardly substitute gas storage completely. According to some estimates (Global Insight, 2004) gas storage provided about 50% of seasonal swing all over Europe, with significant differences within single Countries.[4] To the extent that demand fluctuations mainly concern the household market, gas sellers may conclude interruptible supply contracts with industry and power generation, by offering price discounts to customers that either own multifuel appliances or are anyway willing to bear the risk of supply interruptions. Utilities involved in power generation and supply both to the downstream gas market and the electricity market ("dual fuel market") may accordingly switch multifuel power generation plants to fuel oil or coal in order to maximise gas flows devoted to the temperature sensitive market. Interruptions of industrial customers and fuel switching in power generation plants can be considered a substitute for peak deliverability of storage plants. As a market for flexibility does not exist it is hard to define a value for storage substitutes. One way would be to compare the output value of gas suppliers which use a different bundle of storage substitutes. Another way would be to compare the willingness to pay for storage of gas suppliers differing in their flexibility bundle.

Storage tariffs can be negotiated between storage companies and gas sellers (as in most European Countries and in exceptional cases in the US) or regulated by an independent authority (as in Italy, Spain, Belgium and in most US local markets). Negotiated tariffs allow to discover the value of storage through price signals but the market power of storage suppliers distorts market prices, thus inducing an inefficient allocation of storage resources except when there is enough competition in the storage market. Regulated tariffs may be cost-reflective and effective in controlling the exercise of market power by the storage firm, however they prevent the value of storage from being signalled by prices. Moreover, with congestion the allocation of storage capacity among gas suppliers also depends also on the rationing mechanism included in access rules. Most European Countries adopt "first come first served" methods and allow long term booking of storage capacity. In Italy storage capacity is rationed among gas suppliers according to their market share in the temperature-sensitive market, due to the implementation of public service obligations concerning supplies to households. Auctions are used in a minority of Countries, and for small amounts of storage capacity. To the extent that gas suppliers obtain access to a portion of storage capacity which is not consistent with the value they assign to storage, inefficiency in the allocation of storage capacity ensues. Moreover competition in the downstream market for gas will be affected by distortions in the access to an essential input like storage.

The *Essential Facility* doctrine deals with the imposition on the owner of an essential input of the "duty to deal" with other companies requiring access to the same input. Granting access to these companies prevents abuses of a dominant position by the owner of the essential input, when it operates in the market as a "bottleneck monopolist". Therefore when an asset is considered an essential facility, its owner cannot oppose a "refusal to deal" on the basis of his property rights. The latter are constrained by the need to protect competition in the market where the services granted by the facility are deemed essential.

---

[4] An extensive analysis of the flexibility sources in European Countries before the implementation of market liberalisation is provided in International Energy Agency (2002).

Gas storage assets are implicitly recognized as essential facilities by the last liberalisation directive, to the extent that unbundling and non-discriminatory third party access to storage facilities have already been imposed by the European Commission. However leaving member Countries the option between negotiated and regulated TPA may cast some doubts on the nature of gas storage plants in each national gas market.[5] Negotiated storage tariffs should be implemented in principle when there is a sufficient degree of competition in the storage market.[6] If storage facilities are owned and managed by a bottleneck monopolist, regulated third party access seems to be necessary in order to avoid the exploitation of market power. Otherwise the abuse of the dominant position could materialize in the imposition of excessive storage tariffs, to be sanctioned ex-post only on the basis of the non-discrimination criteria. Therefore once it is ascertained that gas storage is an essential facility, the need to implement regulated third party access ex-ante seems to be a logical consequence.

The essential facility doctrine cannot tell with a sufficient degree of precision and generality which features identify an asset as an essential facility.[7] However the antitrust practice in the US has developed a test to ascertain the economic justification of refusal to deal opposed by the owner of an asset that could potentially be an essential facility (Pitovsky, Patterson and Hooks, 2002). Such a test could be conveniently extended also to the European experience. Castaldo and Nicita (2007). propose a five steps test to identify the requirements to be satisfied: (1) dominant position of the facility owner in downstream markets, (2) unjustified and effective refusal to deal, (3) feasibility of shared access, (4) essentiality of the facility, (5) non-duplicability of the facility. In order to apply the test to storage facilities we examine all five steps that are supposed to be cumulative and, according to Castaldo and Nicita (2007), hierarchically fulfilled.

Dominant position of the facility owner. Dominance should be referred not only to the market where the services of the facility are sold (the storage market in our case) but also to the market position of the facility owner in downstream markets. This first requirement is fulfilled in the case of storage facilities if the storage company has a dominant position not only in the storage market, but also in the downstream market for gas supplies. In most European countries storage is actually provided by a branch of the former integrated utility which also operates in the wholesale market as a dominant supplier of gas. If the facility owner has a dominant position then either a refusal to deal or the imposition of higher access prices may imply a market foreclosure or a raising rivals' costs strategy.

[5] We do not consider this issue at the European market level to the extent that at present a single market for gas is far from being working in the European Union.

[6] It is worth noting that in the US the decision to exempt storage companies from regulated access is left in the hands of an independent regulator, which must assess by a case by case analysis if the degree of market concentration is such as to allow market based rates for storage services.

[7] Areeda, while commenting the diffusion of the doctrine, states in that "The essential facility doctrine is less than a doctrine than an epithet, indicating some exceptions to the right to keep one's creations to oneself, but not telling us what those exceptions are" (Areeda, 1989).

Unjustified and effective refusal to deal. In the case of gas storage this requirement is apparently overcome by the fact that separated TPA to storage facilities is imposed by the second European directive. However the requirement should be intended more broadly to also include cases where access tariffs are excessive and thus such to prevent feasible access by competitors.

Feasibility of shared access. Shared access to the facility should be feasible both from the technical and economic point of view. With respect to this requirement, the degree of rivalry in the use of the facility implied by third party access should be analysed. Once technical feasibility is shown, economic feasibility concerns both the level of access prices and the amount of access that is granted to the facility owner, if the facility is shared with its competitors. Access prices should cover any incremental cost due to sharing, beyond all capital and operating costs. Moreover the owner of the facility should not face any shortage in providing access to its own customers due to the sharing arrangement. In the case of gas storage the technical feasibility of shared access is not an issue. From the economic point of view regulated access can provide for fair access prices to the extent that regulated tariffs are cost-reflective. As to the amount of storage services needed by the supply branch of the dominant firm owning the facility, we can state that in case of storage rationing due to congestion management the final allocation should not impair these needs. However due to the existence of storage substitutes and their concentration in the hands of the dominant supplier, the amount of storage services actually needed by the storage owner should be properly assessed.

Essentiality of the facility. Essentiality means indispensability, i.e. that it is not possible to produce the output for the downstream market without having access to the services provided by the facility. In the case of gas storage the existence of storage substitutes that provide flexibility to gas suppliers may lead to state that the essentiality requirement is not satisfied. However it should also be considered that: (1) the mix between gas storage and its substitutes to provide flexibility affects the value of gas sales in the downstream market; (2) any mix of flexibility inputs must include gas storage, to the extent that it is hard to think that any gas supplier could completely satisfy its own demand for flexibility by just resorting to storage substitutes; (3) when liberalisation starts new entrants in the market for gas supplies are generally not equipped with flexibility substitutes; (4) in some European Countries public service obligations require that gas suppliers are equipped with a due amount of storage capacity, in order to satisfy household demand which accounts for the greatest part of the temperature-sensitive market. While the first point does not concern the "indispensability" of gas storage, the other ones lead to state that the essentiality requirement can be ascertained even in the case of storage facilities.

Non-Duplicability of the Facility. Once the last four conditions are met, the opportunity to duplicate the asset must be considered. Non-duplicability should be assessed both from a technical and an economic point of view. Limitations to the duplicability of storage plants may arise due to the availability of suitable sites. Moreover once a site is found, the availability of new storage capacity may be hampered by the long times required to complete investments. From the economic point of view, gas storage is not a natural monopoly. Any storage plant can supply storage services in competition with other existing plants as minimum efficient scale is generally far from the amount of total storage demand coming from gas suppliers. However the exclusion of natural monopoly is not a sufficient condition to exclude non-duplicability. The latter should be assessed considering the returns on investments in the creation of an alternative facility by a representative competitor. Thus it is necessary to ascertain whether a new entrant is at least able to break-even by replicating the asset. Actually in that case entry cannot be strategically preempted by denying access to the facility. Economic feasibility should however be assessed with respect to market structure. A criterion has been offered by the European Court of Justice with the "Bronner case". If, due to free entry, a symmetric duopoly arises both upstream, in the market for the facilities services, and downstream and furthermore both firms obtain non negative profits, then duplication of the facility is economically feasible (Bergman, 2005). With respect to the to the gas market, it is more frequent to observe a market structure with a dominant firm and a fringe of smaller competitive firms. Therefore, according to the previous criterion, duplication of storage assets seems not to be feasible, considering that market structure is asymmetric both upstream and downstream. Investments in new storage facilities are at present planned all over Europe also by new entrants, showing that infrastructures may be duplicated. According to the long times required to complete new plants, storage can *de facto* remain a transitory bottleneck monopoly in most European markets, until new capacity is delivered to the market by new entrants. The incentives to invest in seasonal gas storage are analysed by Codognet and Glachant (2006) with reference to the UK storage market, where regulation has been removed and storage decisions are supposed to react to spot prices. While investments in storage plants which allow fast and multiple cycles of gas injection and withdrawal (salt cavities) have been common in the UK, investments in huge depleted fields devoted to seasonal gas storage are lacking due to market failures in providing incentives to this kind of investments. Therefore even the market may not deliver an "optimal" amount of investments in storage facilities. Access to existing plants thus remains a regulatory issue.

4.4 The Effect of Storage Rationing on the Productive Efficiency of Gas Suppliers

As shown in Bertoletti, Cavaliere, and Tordi (2008), storage rationing prevents gas suppliers from reaching an optimal mix of flexibility tools. Therefore storage rationing affects the productive efficiency of firms involved in gas sales. In order to concentrate on the cost of flexibility, we assume that the latter is the unique input needed to provide the service of gas suppliers. For the sake of simplicity, we also assume that flexibility must be acquired in the same amount of the final output, indicated by y (this is equivalent to say that the overall production function is of a Leontief-type with respect to flexibility). We also assume that flexibility can be obtained according to a (well-behaved) sub-production function whose intermediate inputs are indicated by the vector $\mathbf{x}$, and that each input x_i has a unit price of w_i. Though a true market for flexibility is hard to define we assume that the specificity of flexibility tools available to each firm can be captured assuming that the unit price w_i of any flexibility input is possibly idiosyncratic to each firm.[8] For instance the advantage of an incumbent can be represented by a lower cost of flexibility vis-à-vis new entrants in the liberalised gas market. Finally, we assume that x_1 is the amount of storage capacity which is procured by the firm in a fixed amount z due to capacity constraints that induce rationing in the storage market. We assume that access to storage is regulated and therefore for all firms the price of storage is fixed and amounts to w_1, corresponding to the regulated (cost-based) tariff for a unit of storage capacity.[9] If there were no restrictions on access to storage the total cost of achieving the amount y of flexibility would be represented by the following cost function:

$$c(\mathbf{w},y) = Min_{\mathbf{x}} \left\{ \mathbf{w}'\mathbf{x} \; s.t. \; f(\mathbf{x}) \geq y \right\}, \tag{4.1}$$

where $f(\mathbf{x})$ is the relevant (sub-) production function. Due to the fact that the amount of storage capacity is fixed, the total cost of flexibility can be represented by a "short run" cost function that we define as the restricted total cost function for flexibility $\hat{c}(\mathbf{w},y,z)$:

$$\hat{c}(\mathbf{w},y,z) = w_1 z + Min_{\mathbf{x}_{-1}} \left\{ \mathbf{w}'_{-1}\mathbf{x}_{-1} \; s.t. \; f(z,\mathbf{x}_{-1}) \geq y \right\}, \tag{4.2}$$

where $\mathbf{x}_{-1}$ represents the vector of all flexibility inputs but storage, whose amount is given by $x_1 = z$. Thus, with obvious notation we can also write the restricted total cost function as follows:

$$\hat{c}(\mathbf{w},y,z) = w_1 z + \hat{c}_{-1}(\mathbf{w},y,z). \tag{4.3}$$

Let $w_1^*(\mathbf{w}_{-1},y,z)$ be the unit price that would induce the firm to demand (conditionally on the output level y and prices $\mathbf{w}_{-1}$) an amount $x_1 = z$ of input one (and the same amount of the other inputs implied by (4.2)) if it were unrestricted, i.e. if there were no rationing of storage but its price were such to lead him to buy exactly the amount obtained when rationing is present. Such implicit price represents the "virtual" *marginal* willingness to pay for that amount of storage and can be defined through the conditional demand function $x(\mathbf{w},y)$ which solves (4.1) as follows:

$$x_1(w_1^*(\mathbf{w}_{-1},y,z),\mathbf{w}_{-1},y) = z. \tag{4.4}$$

[8] From an empirical point of view the specificity of flexibility tools should arise as a difference of the value added to the same amount of gas sold to the same type of customers across suppliers. For example, the flexibility of imports may be coupled with a different price of the commodity and the availability of interruptible contracts is coupled to discounts offered to the industrial customers signing these contracts.

[9] We do not consider here the difference among space, injection capacity and withdrawal capacity. In practice the amount of rationing may be different for the three types of storage capacity offered to customers. For instance rationing might be overcome in terms of space but might persist as far as withdrawal capacity is considered.

Since $\hat{c}_{-1}(\mathbf{w}_{-1},y,z) = c(w_1^*(\mathbf{w}_{-1},y,z),\mathbf{w}_{-1},y) - w_1^*(\mathbf{w}_{-1},y,z)$, we can rewrite the restricted total cost function (4.3) in the following way:

$$\hat{c}(\mathbf{w},y,z) = (w_1 - w_1^*(\mathbf{w}_{-1},y,z))z + c(w_1^*(\mathbf{w}_{-1},y,z),\mathbf{w}_{-1},y). \qquad (4.5)$$

Equation (4.5) allows a simple computation of the (marginal) value of storage when capacity is rationed and access to storage is regulated. In fact, the impact of the availability of a supplementary unit of storage capacity on the restricted cost function is given by:

$$\partial\hat{c}(\mathbf{w},y,z)/\partial z = w_1 - w_1^*(\mathbf{w}_{-1},y,z). \qquad (4.6)$$

Please note that with a virtual willingness to pay for storage greater than the regulated price a marginal increase of storage availability reduces total costs. Therefore $(w_1^*(\mathbf{w}_{-1},y,z) - w_1)$ represents the shadow price of relaxing the constraint on storage availability. This shadow price will be positive only if the firm is actually rationed at $(\mathbf{w},y)$ with respect to storage. On the contrary, the shadow price of relaxing the constraint on storage may be negative, with a willingness to pay less than the regulated price of storage, and implying that alternative flexibility inputs would be more convenient than storage. In this last case a marginal increase in the use of storage would lead to an increase of marginal cost. Therefore we cannot exclude that a gas supplier will prefer to get less storage than the amount he could get on the basis of rationing rules, when the latter allocate storage capacity independently of storage value.

4.5 The Efficient Rationing Rule: An Example

Storage rationing increases total and marginal costs, and induces a sub-optimal demand of the other flexibility inputs. Due to these negative effects on productive efficiency, access to storage might be regulated with the aim of minimizing industry distortions in the allocation of this scarce resource. Accordingly we should find an efficient rationing rule, i.e. a rule that would distribute the rationed input z among firms with the aim of minimizing the total cost of flexibility in the gas industry. A rule that would reach this aim implies that the regulator implements the allocation resulting from the equalization of the shadow costs of storage across firms. In this case the final allocation of storage capacity would be such to respect the heterogeneous values of storage for gas suppliers.

In order to illustrate this point, let us consider an example where flexibility is provided by two inputs, storage (x_1) and a substitute (x_2). The sub-production function for flexibility is a two-input Cobb–Douglas with constant return to scale: $y = \sqrt{x_1 x_2}$. Then we have $c(\mathbf{w},y) = 2y\sqrt{w_1 w_2}$ and $x_i(\mathbf{w},y) = y\sqrt{w_j/w_i}$. As we did in the previous section we assume that $x_1 = z$ and then compute the restricted total cost function $\hat{c}(\mathbf{w},y,z) = w_1 z + w_2 y^2/z$, the restricted conditional demand of the storage substitute $\hat{x}_2(w_2,y,z) = y^2/z$ and the restricted marginal cost function

$\partial\hat{c}(w_2,y,z)/\partial y = 2w_2 y/z$. Therefore in this case the virtual willingness to pay for storage is given by $w_1^*(w_2,y/z) = w_2(y/z)^2$.

With the analysis of the next sections in mind (and without loss of generality for what concerns an illustration of the efficient rationing rule), we assume that the final gas market features two companies: a dominant firm (l), and a single "competitive" follower (f), which can be thought as a competitive fringe of symmetric suppliers. We also assume that the price for the rationed input is regulated to be the unit cost of the storage service ($w_1 = c$), while the price of the unique storage substitute differs across the two firms: $w_{2l} = \alpha w_{2f}$, with $0 < \alpha < 1$, to account for a better access to the storage substitute by the dominant firm. Given the output levels, the efficient rationing rule implies the equalization of the shadow cost of storage for the leader with that for the follower, i.e., $w_{1f}^* = w_{1l}^*$. In the current example this task reduces to the implementation of the following rationing rule:

$$\frac{y_l}{z_l} = \frac{1}{\sqrt{\alpha}}\frac{y_f}{z_f}. \tag{4.7}$$

It is worth noting that the resulting allocation of storage capacity among firms differs in terms of the final allocation of output, as $y_l/y_f > z_l/z_f$. In fact efficiency requires that the firm with the worst access to storage substitutes should be "compensated" with the allocation of a greater proportion of storage capacity than its final sales. Therefore in general neither rules that distribute storage capacity according to final market shares[10] nor "first come, first served" rules lead to cost minimization, as they neglect the asymmetries across firms concerning the availability of storage substitutes. On the contrary, efficient rules should be expected to discriminate among firms and imply asymmetric regulation of access to storage. More capacity should indeed be allocated to firms characterized by higher costs for the storage substitutes.

However, to the extent that the availability (and cost) of storage substitutes is a private and non verifiable information, efficient allocation rules would be very difficult to implement, due to the asymmetric information of the regulator about the technology of each gas supplier, especially considering the idiosyncratic nature of flexibility costs. Clearly, gas suppliers have no incentive to report their amount and/or cost of storage substitutes to the regulator if such a report would reduce the amount of rationed storage capacity allocated to them. In fact, once the competition for market shares is taken into account, any gas supplier even faces the following trade-off: possibly getting more storage capacity than the amount required by cost minimisation leads to an increase of flexibility costs, but to the extent that such an amount is embezzled to the other supplier it may raise rivals' cost and thus increase his own market share. For example, the leader may find it optimal to hoard storage

[10] This kind of rules, satisfying an intuitive fairness criterion, is often used; a possible "equity" justification comes from the practice of it being coupled with public service obligations that require utilities to assure gas sales to households in any event (thus, access rights to storage capacity become proportional to the household market share served by each firm).

capacity in order to increase the follower cost for flexibility.[11] Intuitively, such a strategy should be more profitable to the leader the lower is the regulated price of storage and the higher his cost of storage substitute. The scope for a strategic demand for storage is investigated in the next section. Please note that the regulator may adopt a market mechanisms to elicit firms preferences in terms of storage capacity. For instance, auctions might be used as a suitable rationing mechanism to the extent that bids should naturally depend on the willingness to pay for storage capacity. But even resorting to storage auctions does not eliminate the incentive to hoard storage capacity in order to raise rivals'costs (see next sections). However, with a storage auction the profitability of such a strategy is endogenous to the auction itself as the price paid for storage capacity depends on the bids submitted by gas suppliers.

4.6 Access to Storage with Imperfect Competition in Gas Markets

In this section (and in Sect. 4.7) we will compare the allocation of storage capacities carried out at a regulated price with rationing by some market mechanism (Bertoletti, Cavaliere and Tordi, 2008). Though resorting to a market mechanism is in itself another way to implement third party access to storage by an independent regulator, for sake of simplicity we shall label the first mechanism as "administrative regulation" and the second one as "auctions". Then administrative regulation and auctions are compared considering their effects on competition in the downstream market and on overall social welfare.

We assume that the storage market is a monopoly and that the gas market is characterized by a dominant firm (the market leader) and a fringe of symmetric competitive producers. This last assumption quite correctly reflects the structure of wholesale gas markets in European Countries, which are still dominated by former integrated utilities sharing the market with multiple small new entrants. Often dominant firms in the wholesale markets also own of storage plants, as liberalization directives in the case of storage only require accountancy unbundling.

4.6.1 *The Basic Model*

In this section we will introduce some extreme (though not necessarily unrealistic) assumptions about technology by supposing not only that flexibility is provided by two inputs (storage (x_1) and a unique storage substitute (x_2)), but also that the latter is completely unavailable to the follower. Then while the price of storage (w_1)

[11] In practice if a new entrant has no flexibility tools available but storage, then by hoarding storage capacity the leader can prevent the follower from extending its market share.

is the same for all companies (be it set by the regulator or by an auction mechanism), the price of the other input is w_{2l} for the leader and $w_{2f} = \infty$ for the follower. In addition, the production function is linear: i.e., $y = x_1 + x_2$. Thus the restricted production function of the follower is simply $y_f = z_f$ and his total cost function is given by $c_f(w_1, y_f, z_f) = w_1 z_f$ for any $y_f \leq z_f$, while the restricted cost function of the leader is given by $c_l(\mathbf{w}_l, z_l, y_l) = w_1 z_l + w_{2l}(y_l - z_l)$ for $y_l \geq z_l$. The leader demands storage either because $w_1 < w_{2l}$ and/or just to raise rivals'costs. The main simplification which follows from the linearity of the technology is that the leader's restricted *marginal* cost is unaffected by the amount of storage received (assuming rationing).[12] This simplification has a cost: w^*_{1f} is not well defined (the follower is not really rationed *given* his level of output, while he cannot produce more than z_f) and w^*_{1l} is equal to w_{2l} (so the leader is rationed in terms of storage only if $w_1 < w_{2l}$). As a consequence, an efficient *distribution* of storage just coincides with the only feasible allocation $z_f = y_f$ and $z_l = S - z_f$, where S is the total amount of storage available (we assume that the unit cost of storage c is less than w_{2l}, so that it is never socially efficient to waste some storage). Accordingly, what we investigate in this section is just the strategic behaviour of the leader in presenting his demand for storage.

Concerning the final gas market, we assume that demand is linear: $D(P) = a - P$; as the fringe of competitive followers sells all its feasible output in the market, the residual demand of the dominant firm will be given by $d_l(P) = a - z_f - P$. We assume that $a - 2S > w_{2l}$, so that even a monopolist would be rationed with respect to the available capacity of storage.

4.6.2 *Equilibrium Analysis with Administrative Regulation of Storage Capacity*

In this case we assume that third party access regulation is implemented trough an access tariff and administrative rules governing the allocation of scarce storage capacity. We do not specify the administrative rule chosen by the regulator. It can be one of the rules currently applied in European Countries, which do not consider neither the cost and availability of storage substitutes nor the strategic function that storage may play in order to raise rivals' costs. Therefore in the first period the storage firm distributes storage capacity according to these rules and in the second period firms compete in the gas market on the basis of the amount of storage capacity previously obtained. That is, we assume that the strategic link between the storage market and competition in the downstream market is neglected by the regulator. Be then $w^r_1 = c < w_{2l}$ the regulated price of a unit of storage capacity and γ the percentage of storage capacity assigned to the follower, to give $z_f = \gamma S$ and $z_l = (1 - \gamma)S$, $(0 \leq \gamma \leq 1)$.

[12] In principle, an increase of z_f given the total amount of available storage has possibly both a pro competitive effect (by increasing the supply function of the competitive fringe) and a counter competitive effect by raising the leader marginal cost.

Considering gas market equilibrium, the follower sells in the market an amount of output y_f which is only constrained by storage: therefore $y_f = z_f$. To derive the optimal quantity of gas to be sold by the dominant firm we consider the maximisation of its profit function:

$$Max_{y_l}\Pi = (a - y_l - z_f)y_l - [w_r^r z_l + (y_l - z_l)w_{2l}]. \tag{4.8}$$

From the F.O.C. we obtain the optimal output sold by the leader:

$$y_l = \frac{a - z_f - w_{2l}}{2}, \tag{4.9}$$

the equilibrium output of the industry:

$$y = y_l + y_f = \frac{a + z_f - w_{2l}}{2}, \tag{4.10}$$

and the equilibrium price:

$$P = \frac{a - z_f + w_{2l}}{2}. \tag{4.11}$$

Equilibrium analysis shows that any increase in storage capacity allocated to the follower reduces both the equilibrium quantity of the leader and the final market price in the downstream market. Actually the increase in storage capacity allows the follower to expand his output so that the residual demand faced by the leader decreases, affecting the equilibrium price, which also depends on the leader's marginal cost w_{2l}. Thus, an increase in storage capacity allocated to the follower has a net positive impact on the expansion of the equilibrium output: since $z_f = \gamma S$, we can also note that any increase in the total amount of storage capacity for a given γ induces the same effects just described above. Therefore if a dominant firm is also the owner of storage capacity, then its incentive to invest in storage might be adversely affected by the effects of storage availability to competition in the final gas market. Finally, we can consider what would happen under a "first come, first served" rule, if the leader can choose first his amount of storage. Clearly, it is never optimal for the leader to leave any storage capacity to the follower, because at the very least the leader could choose $z_l = S$, produce the same total output than in the market equilibrium with $z_f > 0$ and get a higher revenue and profit. Thus, in this setting, a leading company "first served" of storage would completely crowd out the competitive fringe (i.e. it would set $\gamma = 0$).

4.6.3 Equilibrium Analysis when Storage Capacity is Auctioned

Access to storage capacity may alternatively be implemented by some market mechanism, whose rules must be set by the regulator. In particular, we suppose that storage capacity is rationed through a *multiunit sealed bid uniform price auction*.

This auction assigns multiple units of storage capacity to each bidder. For each storage unit, bidders must specify their willingness to pay: thus, when bidding for storage, firms present "to the market" a demand function for access to storage capacity. Then the S units of storage are allocated to the S highest bids, but bidders will pay a uniform price p equal to the lowest of the highest bids that are awarded the auctioned units. The storage auction will then establish the unit cost of storage capacity $w_1^a = p$ and this cost can of course exceed the regulated unit tariff for access to storage considered in the previous section $w_1^r = c$. The timing of the model is such that storage capacity is auctioned in the first period and then in the second period firms compete in the gas market. Therefore firms may try to influence the final allocation of storage capacity through their behaviour as bidders in order to affect the result of competition in the final gas market. We continue to assume that these markets are characterized by a dominant firm and a fringe of competitive followers. Based on our assumptions about technology the leader can obtain flexibility not only through storage but also through a storage substitute, while the output of the follower is constrained by the amount of storage obtained in the auction. Then it is rational for the leader to behave strategically also as a bidder for storage and try to manipulate the auction mechanism in order to exert market power in the market for gas supplies.

We are then looking for a sub-game perfect Nash equilibrium and solving the model backwards by considering first the equilibrium in the gas market (second stage) and then the equilibrium values of the auction mechanism (first stage) given the equilibrium of the second stage. As the second stage concerns competition in the gas market, the equilibrium results computed in Sect. 4.6.2 still apply. We can then consider the equilibrium of the auction mechanism. The follower needs storage as an essential input and, given its resulting output and the gas price, demands the amount of storage capacity that maximises his profits:

$$Max_{z_f}\pi = p\left(z_f\right)y_f - w_1^a y_f = \left(\frac{a - z_f + w_{2l}}{2}\right)z_f - w_1^a z_f, \tag{4.12}$$

$$\frac{\partial \pi}{\partial z_f} = \frac{-2z_f + (a + w_{2l} - 2w_1^a)}{2} = 0. \tag{4.13}$$

From the F.O.C. (4.13) we can derive the follower's demand for storage:

$$z_f = \frac{a + w_{2l}}{2} - w_1^a \tag{4.14}$$

(please note that $z_f = S$ if $w_1^a = \frac{a+w_{2l}}{2} - S$). Thus, we can characterize the behaviour of the follower in the storage auction by the following competitive bidding strategy:

$$w_1^a = \frac{a + w_{2l}}{2} - z_f, \tag{4.15}$$

for $0 \leq z_f \leq S$. On the contrary, the leader takes as given the demand for storage of the follower, anticipating that he will get a certain amount of storage capacity at the uniform price set by the auction. Assuming that the storage is rationed ($z_l + z_f = S \leq y$), the leader is led to bid strategically in order to establish the equilibrium price of storage w_1^{a*} within the auction, unless he chooses not to get any storage capacity. Therefore the dominant firm maximises its profit with respect to this price:

$$\begin{aligned} Max_{w_1^a}\Pi &= p\left(z_f\left(w_1^a\right)\right) y_l\left(z_f\left(w_1^a\right)\right) - \\ &\quad \left[(w_1^a - w_{2l}) z_l\left(z_f\left(w_1^a\right)\right) + w_{2l} y_l\left(z_f\left(w_1^a\right)\right)\right] \\ &= \left(\frac{a - z_f\left(w_1^a\right) + w_{2l}}{2}\right)\left(\frac{a - z_f\left(w_1^a\right) - w_{2l}}{2}\right) - \\ &\quad \left[(w_1^a - w_{2l})(S - z_f\left(w_1^a\right)) + w_{2l}\left(\frac{a - z_f\left(w_1^a\right) - w_{2l}}{2}\right)\right] \\ &= \frac{(a - z_f\left(w_1^a\right))^2 - w_{2l}^2}{4} - \\ &\quad \left[(w_1^a - w_{2l})(S - z_f\left(w_1^a\right)) + w_{2l}\left(\frac{a - z_f\left(w_1^a\right) - w_{2l}}{2}\right)\right] \end{aligned} \tag{4.16}$$

Since (given (4.14)) $\frac{dz_f}{dw_1^a} = -1$, from the F.O.C. we can obtain the optimal value w_1^{a*} from the point of view of the market leader as follows:

$$\frac{\partial \Pi}{\partial w_1^a} = \frac{3\left(a + w_{2l}\right) - 6 w_1^a}{4} - S = 0, \tag{4.17}$$

$$w_1^{a*} = \frac{a + w_{2l}}{2} - \frac{2}{3} S \tag{4.18}$$

(in order to establish this equilibrium price the leader just needs to bid w_1^{a*} for the total amount of storage capacity demanded). Please note that $\frac{\partial \Pi_l}{\partial w_1^a} = -S < 0$ if this derivative is evaluated at $w_1^a = \frac{a + w_{2l}}{2}$: thus, it will never be possible for the leader to completely crowd out the competitive fringe if he has to pay a "market" price for storage (this behaviour would be too costly for him in our setting). However, since $w_1^{a*} > \frac{a + w_{2l}}{2} - S$, it will also never be the case that $z_f = S$, i.e. the leader will always engage in some capacity hoarding, irrespective of the value of w_{2l}. Indeed, given (4.21) the follower and the leader will respectively obtain the following amounts of storage capacity:

$$z_f^* = \frac{a + w_{2l}}{2} - w_1^{a*} = \frac{2}{3} S, \tag{4.19}$$

$$z_l^* = S - z_f^* = \frac{1}{3} S. \tag{4.20}$$

Please note that, while it could be expected that the amount of storage optimally obtained by the leader would depend negatively on the value of w_{2l}, and that accordingly corner solutions with $z_f^* = 0$ or $z_l^* = 0$ could possibly arise, in our setting the effect of an increase in w_{2l} is totally offset by the corresponding increase in the follower's bidding strategy (which in turn depends on the increase in the equilibrium price in the gas market) and then in the value of w_1^{a*}: see (4.15) and (4.18). Then, to get the equilibrium price and quantities in the final gas market the previous amounts of storage capacity must be substituted into the equilibrium results of the second stage of the model to obtain the following sub-game perfect Nash Equilibrium:

$$y_f^* = z_f^* = \frac{2}{3}S, \tag{4.21}$$

$$y_l^* = \frac{a - z_f^* - w_{2l}}{2} = \frac{a - w_{2l}}{2} - \frac{1}{3}S, \tag{4.22}$$

$$y^* = y_f^* + y_l^* = \frac{a - w_{2l}}{2} + \frac{1}{3}S, \tag{4.23}$$

$$P^* = \frac{a + w_{2l}}{2} - \frac{1}{3}S. \tag{4.24}$$

These results hold given that storage is rationed and the dominant firm obtains flexibility by partially resorting to a storage substitute. Obviously, the equilibrium output of the industry is increasing with respect to the amount of storage capacity and the equilibrium price is decreasing with respect to the same variable. Moreover, the equilibrium output is increasing (and the equilibrium price is decreasing) with respect to w_{2l}. These results are straightforward, and given the storage allocation results (4.19) and (4.20) come directly from the equilibrium values of the gas market stage (the larger the amount of storage allocated to the follower, the better the performance of the gas market at the final stage: see below for the welfare analysis). In particular, the auction allocates a larger amount of storage to the follower than the administrative rule if and only if $\gamma \geq \hat{\gamma} = 2/3$.[13]

4.7 Welfare Analysis

In order to compare the performance of the administrative regulation of access to storage with storage auctions and draw some conclusions about these regulatory options we have to evaluate the welfare values of the equilibrium outcomes. Social welfare includes consumer surplus in the gas market, and the profits obtained by the market leader (Π), the competitive fringe (π) and the storage company (π_S), i.e.:

$$W = CS + \Pi + \pi + \pi_S. \tag{4.25}$$

[13] In a more general setting, we might expect that the less costly the alternative flexibility to the leader, the larger the amount of storage allocated to the follower in a market solution, the better the gas market performance at the final stage, with a value of $\hat{\gamma}$ decreasing with respect to w_{2l}.

However, assuming that the use of storage is rationed ($S < y$), the only difference between the two allocation mechanisms considered which concerns the storage market comes from the different storage prices $w_1^r = c < w_1^a$. This implies that under the auction mechanism the storage company gets a higher profit, while the overall profit of gas supplier is less, but this is simply a costless transfer from a social welfare point of view. In addition, since overall consumer surplus only depends on the amount of total output produced in the gas market, which in turn only depends on the storage allocated to the follower in the first stage, accordingly we can conclude that the welfare will be the largest under the administrative rule if and only if $\gamma \geq \hat{\gamma}$ (see Appendix).[14] Finally, please note that, under the administrative rule, in our setting an increase of γ will certainly decrease the profit of the leader firm, both as a result of lower gas price and smaller own output, and of greater resort to the storage substitute which is more expensive. On the contrary, the competitive fringe will be positively affected by an increase of γ, since the increase of their gas supply will overcome the negative impact of the decrease in gas price.

4.8 The Regulation of Access to Storage in Italy

Italy is one of the largest gas markets in the European Union. Total gas demand has increased steadily from 2002 to 2007, with gas consumption reaching 84.90 Bscm. Such an increase is mainly due to the power sector, which has become the largest user of natural gas, due to the huge amount of combined cycle gas turbines commissioned in the last decade. With its 10% yearly increase, gas consumption in the power sector in 2007 exceeded 40% of total demand, while residential demand remained below 35%. National gas production has been constantly declining since 1995. At present it covers less than 12% of total demand. Therefore Italy strongly relies upon gas imports accounting for about 87% of national gas consumption.

Due to residential heating consumption, the Italian gas market is affected by strong seasonal fluctuations with the exception of national production which is structurally flat. Seasonal trends are clearly shown in the Fig. 4.1, which illustrates monthly consumption, import, production and stock change patterns over the period 2007–2008.

In 2007 daily summer consumption did not exceed 0.18 Bscm on average, while daily winter consumption amounted to 0,29 Bscm on average, reaching 0.37 Bscm on a peak day Due to inflexible national production and bottlenecks in import capacity (Cavaliere, 2007a) storage provides the main source of flexibility. However storage capacity available to gas suppliers is scarce, due both to huge allocations to precautionary storage and to the lack of new investments. After market liberalisation and despite the steady increase of demand, the amount of storage capacity did not change significantly from the past, when the gas industry was characterized

[14] Please note that there is no question of "cost mix", since in both cases the leader will produce the relevant incremental output $y - S$ by using x_2.

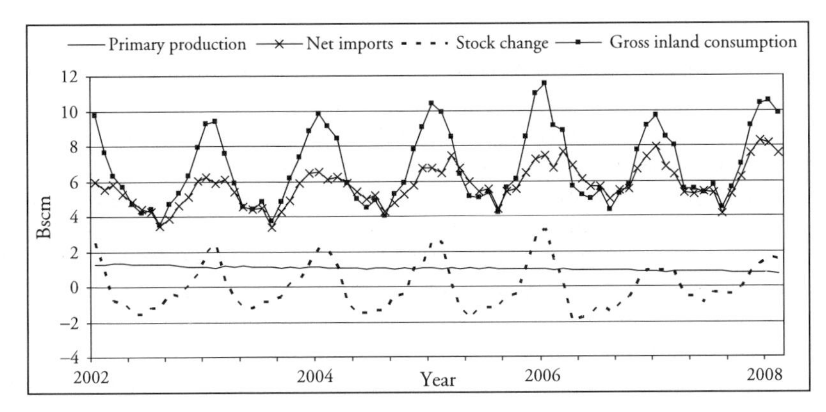

Fig. 4.1 Production, consumption and storage of natural gas storage Italy: 2002–2008. Source: Eurostat

by vertical integration and the monopolization by ENI as a State owned enterprise. In 2007 the ratio of storage capacity to the overall winter and annual consumption was 31,08% and 15,88% respectively. Moreover during the winter period storage plants grant Italy a self sufficiency for about 46 days (Bonacina, Cretì, and Sileo, in press).

When implementing the first European directive (98/30/EC) concerning gas market liberalisation, Italy introduced legal unbundling of storage from other activities of the gas industry and opted for regulated TPA to storage services, going significantly beyond EU requirements that, at the time, did not even require separate access to storage with respect to the transmission network. As a result storage started to be supplied by a new company Stogit, which is completely owned by ENI and account for 98% of storage capacity, consisting in eight depleted production fields. The remaining 2% of capacity was owned by Edison. Furthermore, in 2002 the Energy Regulator (Autorità per l'energia elettrica e il gas, from now on AEEG) provided a cost-based regulation of storage tariffs and provisional rules for access to storage capacity. Stable rules concerning access conditions were provided by resolution 119/05, incorporated into access codes of storage companies with the approval of AEEG. However while scarcity of storage capacity persists, the rationing rules adopted by AEEG did not change significantly with respect to provisional rules adopted in 2002, which were based on pro-quota allocations adjusted to market shares held by gas suppliers.

The fact that Italy opted for regulated access to storage can be evaluated in the light of our previous discussion of the essential facility doctrine, by applying the essential facility test presented in Sect. 4.3. As for the first requirement (*Monopolization* or *Abuse of a dominant position* according to the EU antitrust tradition) it can be recalled that ENI not only is a de facto monopolist in the storage industry, through its control of Stogit, but also continues to act as the dominant firm downstream, in the market for gas supplies (Cavaliere, 2007a). With respect to the second requirement, while *duplication* is technically possible, in practice it still remains ineffective. Actually storage activities are carried out through concession contracts granted by the State, through the Ministry of Industry. Following the implementation of gas market liberalisation, in 2001 the Ministry of Industry published a list of new depleted fields to be transformed into storage plants But the tendering procedure necessary to grant the new storage concessions is still far from being completed. Authorisation procedures are still underway for six new storage sites (five depleted fields and one aquifer facility). At the end of 2006 five more depleted fields were offered by the Ministry of Industry, but tendering procedures are not yet completed. Therefore it may be stated that in Italy institutional procedures to be followed in order to obtain storage concessions work as a barrier to entry, also considering the need for environmental impact assessment and the long time span needed to transform a depleted field into a storage facility. As for the *essentiality* requirement, it can be noted that national production cannot substitute for storage as a supply source, as it is inflexible and more and more negligible. Import flexibility is asymmetrically distributed among gas suppliers and especially benefits the incumbent, due to its historical procurement experience in the international gas market. As to the supply opportunity offered by the spot market, it may be noted that while Italy has a virtual hub for gas exchanges (Punto di Scambio Virtuale, from now on PSV), its liquidity is far from being satisfactory, despite the recent growth of transactions. Interruptible contracts with industrial customers are less and less common though we do not have official data, estimates provided by gas suppliers show that in 2006 interruptible contracts accounted for about 7% of total industrial consumptions. Fuel switching is an opportunity available to some of the main gas suppliers also operating in power production. However the experience of the gas shortage in 2006 showed that logistic problems and the cost of fuel oil storage can prevent fuel switching in dual fuel power plants. Furthermore it is important to note that even the gas suppliers that can rely on storage substitutes cannot avoid access to storage. Storage actually works as a supplier of last resort, especially considering that present it is the main or unique flexibility tool that is used by shippers for balancing their gas flows into the national transmission network. Finally, considering the *feasibility of sharing*, access to existing storage facilities has proved to be technically and economically feasible: during the period 2007–2008 Stogit has provided storage services to 34 customers (AEEG, 2008). However the persistent scarcity of storage capacity still gives rise to congestion issues to be dealt with rationing procedures.

In 2002, at the start of the first regulatory period, storage services were defined by AEEG and a priority ranking for access to storage capacity was established as a basis for storage allocation in case of congestion. Beyond access to storage granted to the transmission system operator in order to assure physical balancing of the high and medium pressure network, storage availability included three different services: a) *mining storage*, designed for the optimization of national production; b) *strategic storage*, concerning precautionary gas stocks devoted to security of supply;[15] c) *storage for seasonal modulation*, designed to respond to the summer–winter

[15] Each importer must have precautionary stocks equal to 10% of the amount of gas imported from non-EU Countries. Both Stogit and Edison Stoccaggio are required to keep gas stocks for this purpose (5.08 Bscm and 0.02 respectively) and the obligation of importers is fulfilled by renting

consumption differential and to ensure supplies during peak winter days. In addition to these regulated services, during the first regulatory period storage companies were allowed to also offer non regulated customized services, mainly consisting in short-run counterflows and storage parking to satisfy commercial needs of gas suppliers and useful to fulfill balancing obligations without paying balancing penalties. Due to the residual and negligible nature of the storage capacity supplied by Edison, the company was allowed to supply just non-regulated customized services. During the second regulatory period customized services were also regulated by AEEG.

In 2002 a regulated cost-based storage tariff was introduced, substituting for the previous tariff set by Stogit just for the period 2000–2001. It must be noted that regulated storage tariffs were not defined separately by Stogit for each depleted field operated (as was done in the UK for instance). Therefore the storage tariff concerns access to the whole storage capacity supplied by Stogit, not to a single storage field. Such a choice was mainly due to the fact that the facilities belonging to each storage company were grouped into a single virtual hub of the national transmission network, when transmission tariffs were regulated according to the entry-exit methodology (Cavaliere, 2007a). One of the shortcomings of such a choice is that cross-subsidies among storage fields are implicitly allowed and if a competitive market for storage were in place, new entrants with a single less efficient field would be at a disadvantage. The regulation of storage tariffs also contained some pro-competitive aims, to the extent that new storage concessions eventually granted during the first regulatory period were exempted from regulation. However new concessions did not materialize due to the very long delays in the authorization process quoted above. At the start of the second regulatory period pro-competitive aims were explicitly abandoned by the regulator in order to concentrate on the expansion of storage capacity. To pursue this aim a higher rate of return is allowed for new investments, beyond the 7.1% rate before tax granted to any capital investment.[16] Furthermore investors are entitled to an exemption from TPA for at least 80% of the new capacity over a period of 20 years, according to the implementation in national legislation of Article 22 of the second European directive on gas market liberalisation (2003/55/EC).

During the second regulatory period the storage services supplied by Edison stoccaggio were also included in the regulation, by setting a single national tariff for access to storage. The motivation for such a regulatory choice was that Edison Stoccaggio would have been at a disadvantage vis-à-vis Stogit, due to the characteristics of its storage fields, that would have led it to set higher tariffs because of higher storage costs. However such a claim appears quite surprising considering that in a market with excess storage demand, Edison would have been able to sell its capacity even at a higher price. The single national tariff, shown in the following Table 4.1 encompasses unit fees for storage space, injection and withdrawal capacity

gas stocks. In addition an obligation is placed on households suppliers to ensure supplies in the case of "1 out 20" winters.

[16] For investment in the development of new storage facilities, the rate of increase is 4% for 16 years, while the expansion of capacity in old concessions is allowed a 4% increase for 8 years.

Table 4.1 Structure of the multipart storage tariff

Unit Fees	Space (Euro per GJ per year)	Injection (Euro per GJ per day)	Withdrawal (Euro per GJ per day)	Movement (Euro per GJ)	Strategic Storage (Euro per GJ per year)
Amount	0.155673	9.503475	11.295975	0.102119	0.156773

Source: AEEG (2006a)

and actual gas movements, plus a specific charge for capacity allocated to strategic storage (including just space).

It is worth noting that the Italian storage tariffs remain the lowest in Europe. This is due not only to the fact that Italy opted for regulated TPA in order to control the exercise of market power by the *de facto* storage monopolist, but also to the particular features of the national storage fields which affect production costs. These fields show a good efficiency in the ratio working gas/cushion gas, which is equal to 65–70%, and a good permeability of the geological formation, allowing high flow rates with a moderate number of wells (Bonacina, Cretì, and Sileo, in press). Moreover the dimension of some storage fields is huge compared to other depleted fields existing all over Europe (Global Insight, 2004), with a positive impact on scale economies. The old economic life of some storage concessions may be such that some assets have already been completely depreciated (Di Renzo and Traini, 2006) All these factors contribute to reduce storage costs and can thus explain the low level of cost-reflective storage tariffs, though the controversial valuation of cushion gas also contributed to this result.

Though cost-reflective tariffs helped to control Stogit's market power, they could not reflect the value of storage for gas suppliers and did not signal the scarcity of storage capacity and the need for new investments. However the lack of storage capacity affecting gas suppliers is due not only to the absence of new investments, but also to the amount of precautionary gas stocks blocked to ensure security of supply and to the rationing procedures followed by AEEG to allocate the existing storage capacity. According to the Ministry of Industry, precautionary gas stocks should amount to 5.1 Bscm. These gas reserves add to 9.4 Bscm of cushion gas and to further stocks of pseudo-working gas amounting to 4.6 Bscm. The latter is a peculiarity of the Italian storage system and is defined as the amount of gas to be steadily kept underground in order to ensure coverage of peak demand even in dramatic (and unfrequent) weather conditions that could occur at the end of the winter season when gas stocks are structurally low. Working gas capacity available for seasonal storage amounted to 8.5 Bscm in 2008. It is worth noting that the criteria that led the Ministry of Industry to set the current amount of precautionary gas stocks are still unclear. To the extent that precautionary gas stocks reduce the amount of capacity available for seasonal storage, a cost-benefit analysis should be necessary to choose the dimensions of precautionary gas stocks, especially considering the persistence of excess demand in the market for seasonal storage. Furthermore even if access to precautionary gas storage is heavily penalized, resorting to these stocks is quite common for gas suppliers in order to fulfill their balancing obligations when they have exhausted the capacity at their disposal for seasonal storage. Though access to precautionary gas stocks is costly and such stocks must be restored after being used, in practice they work as a supplier of last resort, despite the fact that they are supposed to be kept underground to face the risk of interruption of imports from non-EU Countries.

Actually during the period 2005–2006, excess storage demand amounted to 2.2 Bscm, in terms storage space and to about 94 Msmc in terms of peak-day withdrawal (Di Renzo and Traini, 2006). It is worth noting that such rationing of storage demand does not satisfy storage requirements linked to supply of industrial customers and power generation plants. The rationing procedure follows the priority ranking of storage services presented above and is such that seasonal storage capacity is always lower with respect to the requirements of gas suppliers. Therefore the available amount of capacity is allocated on a pro-quota basis, according to the market share held by gas suppliers in the temperature sensitive market (the market concerning households and customers with a yearly consumption of less than 200.000 smc) which is also the market segment protected by public service obligations. The main shortcoming of this rationing criterion is that it completely neglects the asymmetric distribution of flexibility substitutes among gas suppliers, at the risk of allocating more storage capacity than needed to some of them. A recent empirical research (Bonacina, Cretì, and Sileo, in press) confirms such a belief to the extent that a share of the sample considered states that some gas suppliers might have excess capacity but with hold it, instead of selling it in the secondary market.[17] The same research considers the suitability and costs of flexibility instruments, that we report in the following Table 4.2.

As shown by Table 4.2 access to seasonal and precautionary gas storage (i.e. storage with penalties) appears to be most suitable flexibility tool both for seasonal modulation and peak shaving, but when considering cost issues the ranking of

Table 4.2 Flexibility Instruments

Ranking	Suitability for seasonal modulation (decreasing)	Suitability for winter peak (decreasing)	Costs (increasing)
1	Storage	Storage	Storage
2	Contract flexibility	Storage with penalties	Interruptibles (no thermo)
3	Storage with penalties	Spot market (PSV)	Contract Flexibility
4	Spot market (PSV)	Contract flexibility	Interruptibles (thermo)
5	Interruptibles (thermo)	Interruption own thermo	Storage with penalties
6	Interruption own thermo	Interruptibles (thermo)	Interruption own thermo
7	Interruptibles (thermo)	Interruptibles (no thermo)	Production
8	Production	Production	Spot market (PSV)

Source: Bonacina, Cretì, and Sileo (in press)

[17] A problem of capacity hoarding concerning storage is also signalled by the research carried out by the European Commission about competition in the internal gas market (European Commission, 2007).

flexibility tools changes: the most used flexibility inputs are not the cheapest ones. In our opinion such a result highlights a problem of availability of storage substitutes. Moreover the same research confirms that the ranking of flexibility costs varies from gas supplier to gas.supplier. As a matter of fact ENI, the dominant firm, has often required less storage than it was entitled to demand on the basis of its market share in the temperature sensitive market, suggesting that it can rely on flexibility tools less costly than storage.

Therefore one wonders if the adoption of a market based procedure would contribute to lower excess storage demand, to the extent that the willingness to pay declared in the framework of an auction is expected to also depend on the value of storage which in turn depends also on the availability of other flexibility tools. Even if rationing through auctions were not be immune from the incumbent' strategic behaviour, still giving rise to some capacity hoarding, our theoretical results confirm that new entrants could obtain more storage capacity through auctions when the incumbent supplier obtains much more through a regulated pro-quota allocation with a positive impact on social welfare. Moreover the adoption of storage auctions would improve market signals concerning storage scarcity and auction revenues could help to finance the expansion of storage capacity. Even the availability of alternative flexibility tools could increase to the extent that the higher price paid for storage capacity would lead gas suppliers to resort to storage substitutes. The scarce availability of alternative flexibility tools in Italy might also be the result of the low price paid for gas storage.Therefore one wonders how gas suppliers would behave in the case of an increase in the availability of storage capacity coupled with the higher prices that are expected to arise from a market mechanism.

Furthermore the rationing procedure adopted in Italy appeared to be particularly critical during the gas shortage of the winter 2005–2006 (Cavaliere, 2007b). At that time a lot of gas suppliers resorted to their storage capacity to withdraw gas in order to produce electricity which was then exported to France, and exploited the sudden spread between the French and the Italian power exchange, despite the fact that this storage capacity ought to be in principle devoted to the household market. Also due to the occurrence of the coldest winter in the last 20 years, the summer–winter consumption differential was partially covered with resort to precautionary gas stocks for an amount of 1.2 Bscm. In order to avoid any risk of interruption to customers protected by public service obligations the Government had to organize an auction to grant subsidies to industrial customers accepting further interruption of gas supplies beyond those of standard interruptible contracts, with further cost for final customers. In order to avoid further risks during the winter season, the Energy Regulator, not only applied sanctions to companies that used gas in storage to speculate in the electricity market, but also introduced more obligations to be full filled by gas suppliers, like the restocking and upkeep of the reserves in storage and the maximisation of import flows during the winter season. Thus at present the storage market appears to be far from being market driven.

4.9 Conclusions

In vertically integrated industries gas storage is an important optimisation tool for gas suppliers. In liberalised gas markets storage not only arises as a potentially independent industrial activity but also plays a strategic role to the extent that storage availability affects competition in the downstream market for gas supplies. In this Chapter we have tried to provide a first analysis of regulatory issues concerning storage, considering both the productive efficiency of gas suppliers who demand storage as a flexibility tool and the allocative efficiency issues in gas markets that are far from being competitive. Due to the scarcity of storage resources in many European Countries, we have been led to consider efficient rationing rules as an important part of the current regulatory issues.

In this Chapter we have considered neither the issue of ownership unbundling of storage facilities from gas supply nor the incentives to invest in new storage facilities. However, our conclusions about the strategic use of storage by the incumbent are reinforced when the incumbent is also the owner of storage facilities. In that case not only the incumbent can raise rivals' cost by hoarding storage capacity, but it can also prevent an expansion of the follower market share by controlling the pace of storage investments until new entrants have developed their own storage capacity. Without ownership unbundling even the positive effect of a market mechanism on the adverse incentives of the incumbent is diluted. In fact the market clearing price paid by the incumbent gas supplier to the storage company just becomes a transfer price inside the same holding company. We have also neglected the existence of secondary markets for storage capacity, where suppliers could get rents from selling excess capacity even if market prices may approximate the real values of storage. However, if secondary markets were considered the possibility of strategic behavior should be analysed also within these markets, especially considering the opportunity of withholding capacity instead of selling it.

A further limitation of our analysis is that it considers the effect of storage rationing on productive efficiency separately from its effect on allocative efficiency, so that when the incumbent supplier's strategic behavior is analysed its effect on productive efficiency is neglected. If the incumbent gas supplier is also the most efficient one from a productive efficiency point of view – which is rather likely – any increase in the output he supplies would have a positive effect on welfare through an increase of the overall productive efficiency. However this last effect cannot be accounted for in our analysis. Such a limitation is due to our assumptions about the technology of gas supply. At present we do not have empirical evidence about the technology of flexibility and consequently our assumption about a linear production function is just for the sake of simplicity. But considering the issues of both productive and allocative efficiency would also allow to discuss the question of the efficiency of entry. Such an issue has been put aside in the debate about the liberalisation of the European gas market, to the extent that the unsatisfactory results about competition probably led to welcome any additional entry or any additional expansion of new entrants. On the contrary, the issue has been widely considered in the telecommunication markets, also with reference to the regulation of access prices, and it should deserve some analysis also in the gas market.

4.10 Appendix

First we derive consumer surplus (CS_R) when access to storage is rationed by the regulator:

$$CS_R = \left[\left(\frac{a-1}{2} \right)^2 + \gamma S \left(\frac{a-1}{2} \right) + \left(\frac{1}{2} \gamma S \right)^2 \right] \frac{1}{2} \tag{4.26}$$

Considering instead the equilibrium outcome arising from the adoption of a market mechanism we can derive consumer surplus (CS_A) when storage capacity is auctioned:

$$CS_A = \left[\left(\frac{a-1}{2} \right)^2 + \frac{2}{3} S \left(\frac{a-1}{2} \right) + \frac{1}{9} S^2 \right] \frac{1}{2} \tag{4.27}$$

By inspection it is easy to check that these two expressions are identical when $\gamma = 2/3$. Actually it is easy to check that the equilibrium price and the output sold in the gas market are identical in both regulatory systems if $\gamma = 2/3$. Therefore consumers would be indifferent to the regulation of access to storage if regulatory rules were such as to allocate storage capacity exactly as auctions do. Storage auctions are then optimal from the consumers point of view, except when the regulatory rules are such to allocate more storage capacity to the competitive fringe ($\gamma > 2/3$).

Let us then analyse the impact on social welfare, considering first the case of administrative regulation:

$$\begin{aligned} W_R = CS_R + \left(\frac{a-1}{2} \right)^2 - \gamma S \left(\frac{a-1}{2} \right) + \left(\frac{1}{2} \gamma S \right)^2 + \\ + S(1 - w_{r1})(1-\gamma) + \left(\frac{a+1}{2} - \frac{\gamma S}{2} - w_{r1} \right) + S(w_{r1} - c) \end{aligned} \tag{4.28}$$

and then the case of storage auctions:

$$\begin{aligned} W_A = CS_A + \left(\frac{a-1}{2} \right)^2 - S \left(\frac{a-1}{2} \right) + \frac{3}{9} S^2 + \\ + \frac{2}{9} S^2 + \left(\frac{a+1}{2} - \frac{2}{3} S - c \right) S \end{aligned} \tag{4.29}$$

In this case too it is easy to check that $W_R = W_A$ if $\gamma = 2/3$. Therefore, even considering the social welfare criterion, we can state that storage auctions dominate the administrative allocation, except when the regulatory rule is such as to allocate more storage capacity to the competitive fringe ($\gamma > 2/3$).

References

Bergman, M. A. (2005). When should an incumbent be obliged to share its infrastructure with an entrant under the general competition rules? *Journal of Industry, Competition and Trade, 5*, 5–26.

Bertoletti, P., Cavaliere, A., and Tordi, A. (2008). The regulation of access to gas storage with capacity constraints. *mimeo*.

Bonacina, M., Cretì, A., and Sileo M. (in press). Gas storage services and regulation in Italy: A delphi analysis. *Energy Policy*.

Castaldo, A. and Nicita, A. (2007). Essential facility access in Europe: Building a test for antitrust policy. *Review of Law and Economics, 3*(1).

Cavaliere, A. (2007a). The liberalization of natural gas markets: Regulatory reform and competition failures in Italy. Oxford Institute of Energy Studies, Working Paper NG20, currently available at http://www.oxfordenergy.org/gasresearch.php.

Cavaliere, A. (2007b). Liberalizzazioni e Accesso alle Essential Facilities: Regolamentazione e Concorrenza nello Stoccaggio di Gas Naturale. *Politica Economica, 1/2007*, 29–64.

Chaton, C., Cretì, A., and Villeneuve, B. (in press). The economics of seasonal gas storage. *Energy Policy*.

Codognet, M. K. and Glachant, J. M. (2006). Weak investment incentives in new gas storage in the United Kingdom? *mimeo,* available at www.grjm.net/documents/M-K-Codognet.

Di Renzo, A. and Traini, S. (2006). Lo Stoccaggio del Gas Naturale in Italia: Regolazione, Mercato e Criticità. Quaderni di Ricerca ref., ref.Ricerche e Consulenze per l'Economia e la Finanza, Milano.

Durand-Viel, L. (2007). Strategic Storage and Market Power in the Natural Gas Market, available at http://teaching.coll.mpg.de/econwork/Durand.pdf.

European Commission (2007). DG Competition Report on Energy Sector Brussels, SEC (2006) 1724, available at ec.europa.eu/comm/competition/sectors/energy/inquiry/.

ERGEG (2006). ERGEG final Report on Monitoring the Implementation of the Guidelines for Good TPA Practice for Storage System Operators Ref: EO6-GFC-20-03, available at www.ceer-eu.org.

FERC (2005). Rate Regulation of Certain Underground Storage Facilities available at www.ferc.gov/whats-new.

Global Insight (2004). Gas Storage in Europe 2003-2004 Research Report.

Hoffler, F. and Kubler, M. (2006). Demand for storage of natural gas in North-Western Europe. Trends 2005–2030 Max-Planck-Institute Working Paper Series, available at www.ssrn.com/link/Max-Planck-Institute.

International Energy Agency (2002). *Flexibility in Natural Gas Supply and Demand*, available at www.iea.org.

Pitovsky, R., Patterson, D., and Hooks, J. (2002). The essential facility doctrine under U.S. antitrust law. *Antitrust Law Journal, 70*, 443–462.

Chapter 5
Gas Storage and Security of Supply

Anna Cretì and Bertrand Villeneuve

5.1 Introduction

Natural gas consumption has seen a fast growth in the European Union over the last decades. It is challenging the supremacy of oil as the leading source of energy and has reached a dominant position in electricity generation. In 2005, about one quarter of the EU primary energy consumption was based on natural gas, and imports from neighboring producers, mainly Russia, accounted for 35% of the total EU25 demand (DG TREN, 2006a). Dependency on external supplies is going to increase in the next years, as gas consumption in Europe is expected to grow whereas indigenous sources are forecasted to slow down. Including the new member countries, the European dependence rate for gas will amount to 62% in 2010, 81% in 2020 and 84% in 2030 (DG TREN, 2006b). Even though up till now there have been no major interruptions of gas supplies in the European Union, the increasing import dependence raises serious concerns about security of gas supply. Therefore, strategies against potential disruption are becoming of crucial importance in Europe. The gas disruption that hit Western and Central Europe in January 2006 illustrates the reality of the threats. Ukraine apparently withheld gas as a result of conflictual relationship with the main Russian producer, causing a significant reduction of the gas volume that posed serious problems in the stock management policies of importer countries. The impact was exacerbated by a very cold winter.

A. Cretì
IEFE (Centre of Research on Energy and Environmental Economics and Policy), and Bocconi University, Department of Economics, via Roentgen, 1, 20136 Milan, Italy
e-mail: anna.creti@unibocconi.it

B. Villeneuve
Université de Tours, CREST (Paris) and Laboratoire de Finance des Marchés d'Énergie, CREST – J320, 15 boulevard Gabriel Péri, 92245 Malakoff, France
e-mail: bertrand.villeneuve@ensae.fr

A. Cretì (ed.), *The Economics of Natural Gas Storage: A European Perspective*,
DOI: 10.1007/978-3-540-79407-3_5,

By financing pipeline construction, a diversified portfolio of long-term contracts with producers is the primary supply insurance against supply interruptions. Security of supply targets can also be met by increasing system flexibility (fuel switching, interruptible contracts and liquid spot markets). However, these mechanisms have a limited capacity to absorb shocks such as severe weather, technical breakdown, terrorism, which would endanger all the European countries at the same time and trigger a supply crisis (Weisser, 2007). In the short-medium term, to ensure uninterrupted services when events of "low probability but high potential market impact" (Stern, 2004) hit the system, precautionary gas storage is indispensable.

The conditions to be fulfilled in relation to security of supply and availability of storage for existing suppliers and entrants have been specified by national laws in application of the Directive 98/30/EC on the liberalization of the gas market. The European discipline has continued to stress the matter of security of supply both in the Directive 2003/55/EC, fostering competition in gas markets, and in Directive 2004/67/EC. The latter has obliged European countries to define the roles and responsibilities of all market players in ensuring gas availability and set minimum targets for gas storage, at national or industry level. The storage policy has to be transparent, and member states have to publish regular reports on emergency mechanisms and the levels of gas in storage that the Commission will monitor – a procedure which to date is in place in the US only.

To implement these Directive, some countries have set specific storage obligations. In Italy, entrants importing non-EU gas are required to hold stocks equivalent to 10% of the annual supply. In Spain, overall gas supply dependence upon any single external supply source must not exceed 60% and gas companies are obliged to keep gas reserves of at least 35 days of supply. In Denmark, the integrated gas firm has designed its back-up and storage capacity to be able to continue supplies to the non-interruptible market in case of a disruption of one of the two offshore pipelines supplying gas to the country. In France, strategic stocks can withstand disruption of the largest source of supply up to 1 year. UK, instead, has decided not to take specific measures to safeguard security of gas supply. The Department of Trade and Industry in the UK has stated in 2007 that existing regulatory measures could be relied on to transpose virtually all of the mandatory parts of the Directive 2004/67/EC. This case seems to us emblematic and worth investigating in this Chapter on security of gas supply.

So far, the economic literature has not addressed the medium – term security of supply problems that European countries face when the gas market is concerned. As we argue in Sect. 5.2, either the existing models ignore the existence of long-term contracts, therefore focusing on extraction rate of producer countries when there is a trade-off between present and future security, or they look at cartelized supply. Those are not the primary issues in managing secure gas services to Europe. Our model fills this gap by explicitly addressing the incentive to store by a private sector which considers the probability of a supply disruption. Private stockholding decisions balance the valorization of gas in the event of a crisis with its carrying costs (capital immobilization and technical costs).

The issue is a very complex one, so simplification is essential if any progress is to be made. We assume that the size of disruption is single-valued and known, its probability is also known and stationary, and disruption marks a permanent transition to a state of lower excess supply. Given these assumptions, we derive the dynamics of accumulation and drawdown in a continuous time context. Section 5.4 presents the dynamic model and its main assumptions, whereas the following Sect. 5.5 provides the characterization of the competitive equilibrium. We find equilibrium prices, optimal target stocks and drainage time.

Security of gas supply has a crucial policy dimension. To address this important issue, we provide a complete theoretical treatment of the effects of public interventions and provide a method to evaluate antispeculative storage policies in a dynamic setting (Sect. 5.6). In the last part of the Chapter (Sect. 5.3), we evaluate the cost of imperfect policies in a detailed example calibrated on UK data. "Is the UK right?" is one of the questions we analyze. We provide suggestions as to the benefits of stock targets and management rules to be decided at the country level. We find that there is a rationale for accumulating strategic gas stocks, thus casting some doubts on the current UK gas security policy. In the Appendix, we also suggest two important extensions of the basic model that deal with specific characteristics of the gas industry: non negligible injection and release costs, and limited storage capacity. The main findings of our model are shown to be robust to the modified settings.

5.2 The Issue of Security of Supply in the Literature

Since a decade ago, gas security was not a big issue; therefore, the theoretical literature on energy supply security has mostly been inspired by the question of oil as a precautionary reserve. There are two sets of models, mostly inspired by the theory of exhaustible resources: works that consider the extraction rate of one country when foreign import, though needed to complement national production, can suddenly default, and those that introduce strategic behavior of consuming countries confronting oligopolistic or cartelized supply.

The first group of models shows that generally, with imports subject to disruption, an importing country faces a trade-off between current and future security of supply. Two effects are at stake: on the one hand, inasmuch as foreign supply substitutes cannot expand quickly in a disruption, there is a motive for speeding up domestic production to reduce near-term economic losses; on the other, the scarcity value of domestic reserves is increased, providing an incentive for conservation to anticipate future emergencies. This trade-off has been analyzed by several authors (for example Stiglitz, 1977; Sweeney, 1977; Tolley and Wilman, 1977; Hillman and Van Long, 1983; Hughes, 1984; see also the Introduction of this book).

Given the extent of EU import dependence and the unavoidably declining rate of internal gas production, we argue that those models focusing on the extraction rate of a domestic resource do not provide an adequate answer to the actual problems of short-medium term security of gas supplies. For this reason, we focus on the occurrence of two states, abundance and crisis, that are likely to happen in the medium – term, in which both the seasonality of demand and the exhaustibility of gas can be practically neglected. Moreover, in our model, there is no distinction between domestic and foreign production.

The second group of models includes the analysis of strategic interaction among importing countries and exporters (Nichols and Zeckhauser, 1977; Crawford, Sobel, and Takahashi, 1984; Devarajan and Weiner, 1989; Hogan, 1983). These work stress that since the oil stockpiling strategy can be considered as a public good, free-riding problems can be solved by policy coordination at the supra-national level. This kind of action is today a matter of fact in the gas market: the European security of supply measures are a good example in this direction. Moreover, long-term gas contracts seem to have adequately tackled the commercial and political issues between exporters and importers. Finally, security of gas supply is probably not a question of strategic interactions against a cartelized group of producers, as Russia, Norway, Algeria do not form a cartel.

None of the previously cited models addresses the question of how to reach any desired stock level and how to deal with uncertainty about the duration of the supply disruption. Efficiency loss of recommended storage policies may be underestimated. These aspects are of crucial importance in the European context. In the most significant attempt to address these questions, Teisberg (1981) developed a macroeconomic dynamic programming model that accounted for by minimizing a US cost insecurity function due to oil import disruption.

Our model shares with Teisberg's one the stochastic specification for the supply disruption. However, we put forth a rather different perspective, since our model focuses on a microeconomic rationale (i.e. arbitrage theory) to explain stocks formation and drawdawn. Therefore, we emphasize not only the stockpile pattern, but also the (endogenous) price dynamics. This is also a noticeable difference with respect to Bergström, Glenn, and Mats, (1985) in which stocks are built up at the exogenous world price as they analyze the case of a "small country" that does not influence the international trade of the commodity exposed to an embargo threat.

Teisberg also assumes that in a finite time horizon the price path will hit a predefined back-stop value at which market insecurity is no longer a problem and remaining stocks will be sold. This hypothesis, appropriate in a long-term perspective, is not well suited to short-medium term crisis management that our work considers. In this respect, we show a key result: the optimal limit stock that storers will built by foreseeing the crisis is never really attained. In fact, stockpiling before the disruption increases gas prices, so accumulation is all the faster in so far as potential profits loom large. The limiting factor to accumulation is that the value of the stored cubic meter in case of crisis decreases as stocks pile up. This prevents the economy to reach in finite time the target stock.

Finally, on the policy dimension, we do not explicitly consider tariff or quotas, as this kind of public intervention does not seem realistic in the European gas market. The understanding of potential market failures or imperfections is of crucial importance in the perspective of the European Directive aimed at improving the security of gas supply. Dealing with extraordinary circumstances involves a public intervention role and requires consistency of the public approach both before and after an extraordinary circumstance. For example, stockholders may fear antispeculative measures taken once the crisis has occurred. We show that this lack of protection of property rights is likely to discourage storage completely, and that responsible policy consists in a series of measures (subsidies, public agency) taken ex ante.

This kind of analysis resembles the Wright's and Williams's (1982) approach. These authors emphasize the role of public storage in managing oil import disruption in a stochastic economy and its relationship with private stockpiling. However, the assumption of i.i.d. shocks used by Williams and Wright cannot capture the persistence that supply crises caused by "low-probability-high impact event" are likely to exhibit. We are able to analyze not only the incentives to manage stock by a public agency and how to fund it, but also the impact of forced accumulation and drainage rates representing a typical imperfection in policy measures taken to implement security of gas supply. In particular, we evaluate the welfare loss of second best policies by a parsimonious method that does not rely on estimates of the cost insecurity function, as it has been suggested by some of the surveyed models, but only requires knowledge of the price functions corresponding to specific policy interventions, the supply and demand fundamentals, and the stochastic process.

5.3 Storage and Security of Supply in the UK

The British gas market is very mature and it is characterized by strong gas penetration in all end-user sectors. It is worth recalling the supremacy of gas in the UK energy mix: in 2006 the share of gas in the energy mix was 41.6%. Due to environmental constraints (i.e. carbon dioxide emissions reduction policies), investments in combined cycles gas turbines (CCGT), poor penetration of renewable sources and uncertainty regarding the future of nuclear energy (i.e. decommissioned capacity not being replaced as from 2018/2020)[1] the current dominance is expected to expand even further. According to Global Insight forecasts, the gas share in the UK energy mix will approach 67.5% by 2030, thus becoming the primary fuel in the generation mix.

Although the growth rate in national gas demand is unlikely to be dramatic, excluding outliers due to unexpected and exceptional events (i.e. climatic emergency and fire at Rough),[2] DTI has assessed an increase by 1% per year in the years to come. However, patterns diverge on a sector basis. Industrial and residential consumptions have almost stabilized. Progressing steadily along the pattern started in the 1970s, currently the transport sector accounts for more than 40% of the demand for gas by final users. Similar trends apply to power generation.

[1] Source: National Grid website.

[2] The collapse in national gas demand during the winter of 2005–2006 has resulted in a reversion of final energy consumption patterns. Since 2005 annual (monthly) demand for gas has returned below 100 Bscm (12 Bscm), with an average reduction by 1.1% year-over-year in the years 2002–2007.

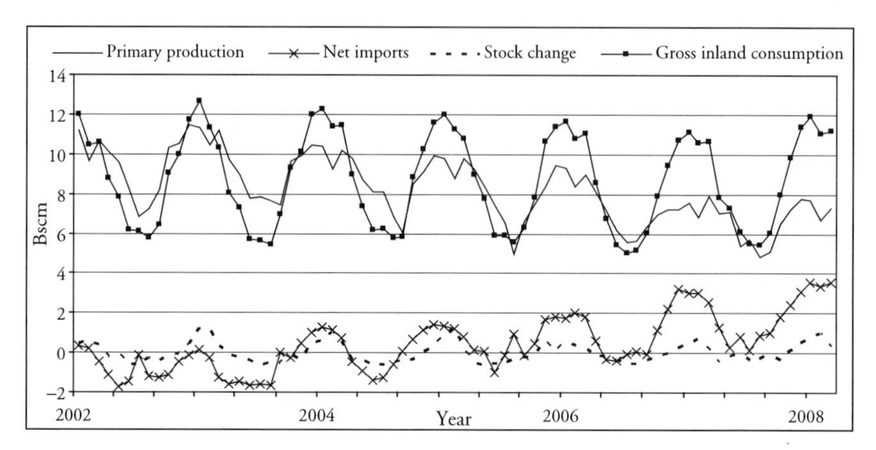

Fig. 5.1 Production, consumption and storage of natural gas in the UK: 2002–2008. Source: Eurostat

Figure 5.1 provides an overview of monthly consumption, net import (i.e. import minus exports), production and stock change patterns over the period 2002–2008. It is straightforward to note that gas market characteristics vary considerably on a seasonal basis mainly due to demand for heating. In the years 2006–2007, average daily demand in the winter has approached 0.33 and reached 0.44 Bscm on a peak-day while the summer consumption values have been 0.24 and 0.36 Bscm, respectively. Seasonal patterns on the demand side do not always match with adequate flexibility on the supply side of the gas market. The traditional source of balancing in the UK has been national gas production which, however, has been declining. In the years 2002–2007, UK natural gas production has been reduced by 37%. According to National Grid, the persistence of recent trends would further halve national production by 2020. The pattern is confirmed by net import developments. According to Fig. 5.1, by 2002 UK natural gas exports have regularly decreased. In 2006 the UK has become a net importer of gas. Henceforth imports have become the secondary instrument for balancing (especially seasonal modulation). In 2007 imports (via both pipeline and liquefied natural gas) accounted for some 19% of UK gas requirements (doubling the share of 2006). Although the portfolio of exporters is quite diversified (it includes Norway-64%, Belgium-16.5%, Algeria-8.5%, Egypt-5%, the Netherlands-4%, as well as other non-EU Countries), security concerns remain at the forefront. LNG imports in the UK have to compete with alternative terminals in both the United States and Europe.

5.3.1 The Gas Storage System

The traditional availability of the high swing at the beach in the UK has lessened the concerns on security of supply (there is no strategic reserve obligation).

Table 5.1 Storage facilities, capacity and ownership in 2007

Facility name	Owner	Space (Bscm)	Deliverability (mscm per day)	Injectability (mscm per day)
Rough	Centrica Storage Ltd.	3.07	49	15
Hornsea	SSE	0.32	18	2
Hatfield moor	Scottish power	0.12	2	2
Humbly grove	Star energy	0.29	7	8
Hole house	EMGS	0.04	6	9

Source: DTI, 2007

Consequently, to date, relative little storage has been built in the UK and the operation of the existing scarce capacity is barely regulated. In the years 2002–2007, the overall working capacity, 4.1 Bscm, has not exceeded some 4% of annual gas demand (7.5% of seasonal modulation requirements according to National Grid). The market-oriented development of the UK storage system has not only affected the extent and operation of the sector but it has also governed the selection of the sites in which gas is stored. The technical framework is dominated by depleted fields and salt cavities at high deliverability rates. It is extensively recognized that such sites are particularly suitable for short-run (peak shaving) purposes.

Notwithstanding the advances in the liberalization process, the separation of transportation and storage activities, and the ending of the former monopolistic market structure, the UK storage system remains highly concentrated on both geographic and operational extents. There are five active storage sites in the UK, each of them being managed by a different Storage System Operator (henceforth SSO). Facilities are really heterogeneous with respect to both the storage capacity they make available and the range of services (medium versus long-term balancing) they can offer. By owning and operating the sole depleted field, Rough, Centrica Storage Ltd[3] manages some 81% of UK storage space, 60% of daily deliverability and 42% of daily injectability. The remaining capacity is shared between Scottish and Southern Energy (SSE), Energy Merchants Gas Storage (EMGS), Scottish Power and Star Energy Ltd, which control (medium-run) salt caverns at Hornsea, Hole House Farm, Hatfield Moor and Humbly Grove respectively (see Table 5.1).

Notwithstanding the persistence of high concentration levels in the storage capacity, since the unbundling of British Gas, according to DTI and the UK gas regulator, market power that could be exercized by Centrica does not seem to be a relevant issue.[4] Furthermore with the commissioning of new storage facilities and the entering

[3] Centrica Storage Limited has acquired Rough storage facility from BGS in 2002.

[4] Natural gas transmission and storage in Great Britain has evolved from an integrated system operated by British Gas to a market with multiple participants. In 1998, Ofgas (Ofgem, office of the gas and electricity markets, is the successor to Ofgas, office of gas supply) recognized that "British Gas Storage was capable of exercising significant market power" and that "such market power was indeed being exercised in ways that were hindering the development of competition." In 1999 Ofgas has issued a decision document on storage deregulation, the "Review of the supply

Table 5.2 New storage capacity

Facility name	Owner	Awaited space (Bscm)	Time
Fleetwood	Canatxx	1.20	by 2012
Bains	Centrica/GdF/First Oil	0.57	by 2011
British Salt	British Salt	1.00	by 2010
Hewett	ENI/Perenco	5.00	by 2015
Hole House	EDF Trading	0.06	by 2010
Isle of Portland	Portland Gas	1.00	by 2015
Holform (ex-Byley)	E.ON. UK	0.17	by 2010
Whitehill Farm	E.ON. UK	0.42	by 2012
Stublach	GdF Storage UK/Ineos	0.40	by 2018
Humbly Grove	Petronas	0.06	by 2009
Albury	Petronas	0.90	by 2012
Welton/Scampton North	Petronas	0.45	by 2010
Bletchingley	Petronas	0.85	by 2015
Esmond/Gordon	Petronas/EnCore	4.10	by 2012
Aldbrough	SSE/Statoil	0.84	by 2012
Gateway	Stag Energy	1.14	by 2012
Saltfleetby	Wingas	0.75	by 2013
Caythorpe	Warwick Energy	0.21	by 2010

Source: Gas Storage Database, 2008

of new market players, market power concerns will become even less serious in the years to come. In fact, if every planned storage infrastructure will enter into operation as scheduled in Gas Storage Database, some 50% and 140% increase in British storage capacity is expected by 2010 and 2015, respectively (see Table 5.2). Except for at most some 10%, such additional capacity will be managed by newcomers. New investments are expected to leave Centrica Storage Ltd with less than 30% of UK storage space by 2013. Finally, by 2015 the former seven companies will manage some 81% of UK storage space.

Concerning the operation of storage facilities, unlike in the Continent, each SSO in the UK sets its injection/withdrawal strategy on a commercial basis, in such a way that the value of its gas stock is maximized. Hence storage services may be either used by SSOs for their internal portfolio and/or sold to shippers. Hence storage facilities in the UK are owned and operated by arbitrage-oriented and profit maximizing market players.

With respect to regulation of storage services, we would preliminary recall that neither the management nor the ownership of gas storage facilities is a licensed activity in the UK. The Office of the Gas and Electricity Markets (Ofgem), the Office

of gas storage and related services," according to which the underground storage facilities owned by British Gas were spun off to a new affiliate, British Gas Storage (BGS), and were removed from the British Gas transporter license.

of Fair Trading (OFT), the Department of Trade and Industry (DTI), the Health and Safety Executive (HSE) and the Competition Commission (CC) supervise the storage activity. Ofgem has a specific role in monitoring the behaviour of SSOs, taking action against their potential discriminatory or anti-competitive practices, deciding on applications for exemption from Third Party Access, arbitrating any dispute over access terms and supervising the negotiation of Standard Service Contracts (SSC).[5] OFT has Concurrent Competition Act powers, and monitors the SSOs pursuant to section 88 of the Fair Trading Act 1973. The other parties (i.e. DTI, HSE and CC) are responsible for transposing European Directives in UK legislation, ensuring compliance with the relevant health and safety legislation, and ensuring fair competition. Despite the amount of bodies supervising the industry and probably thanks to the fair operation of the British gas market, storage facilities are currently exposed to mild regulation.

For the sake of exhaustiveness let's recall the key steps which have led to the present legislative frameworks. The entering into force of the Gas Directives 98/30/EC and 2003/55/EC concerning common rules for the internal market in natural gas, has yielded several changes to the legal and regulatory framework applying to British storage facilities. The transposition of the First Gas Directive[6] into UK law has resulted in the application of mild regulated and negotiated Third Party Access (rTPA and nTPA) arrangements to storage services. Exemptions have been authorized by Ofgem on a case by case basis.[7] The regime lasted few years. The implementation of the Second Gas Directive, by narrowing the application of 98/30/EC, introduced several amendments to both exemption regimes and TPA requirements.[8] Currently the UK is the sole EU country where (mild) rTPA, nTPA and exemption regimes coexist. Mild rTPA is required for Rough and Hornsea which are

---

[5] Ofgem can veto any proposed changes to the SSC.

[6] The First Gas Directive has been transposed into UK law by the Gas Regulation 2000. The latter has integrated the Gas Act 1986 with sections 19A to 19D.

[7] Ofgem may authorize exemptions if the requirements of TPA arrangements were already met and/or if the use of the facility was not necessary for the operation of an economically efficient gas market.

[8] First, to put in place a regulated TPA (rTPA) regime, integrations to the Gas Act 1986 (i.e. sections 19A to 19D) have been amended. Second, a distinction in the exemption regime between existing and new (or upgraded) storage facilities has been set up. In particular existing facilities may be TPA exempted if and only if the use of the facility by other person is not necessary for the operation of an economically efficient gas market while new facilities receive the same treatment if either (a) the use by other person is not necessary for the operation of an economically efficient gas market, or the six exemption requirements contained in section 19A(8) of the Gas Act are met (i.e. the facility or the significant increase in its capacity will promote security of supply; the level of risk is such that the investment to construct the facility or to modify the facility to provide for a significant increase in its capacity would not be would not have been made without the exemption; the facility is or is to be owned by a person other than the gas transporter who operates or will operate the pipeline system connected or to be connected to the facility; charges will be levied on users of the facility or the increase in its capacity; the exemption will not be detrimental to competition, the operation of an economically efficient gas market or the efficient functioning of the pipeline system connected or to be connected to the facility; and the Commission of the European Communities is or will be content with the exemption.

Table 5.3 Rough and Hornsea storage facilities

Facility name	Type	Standard bundled unit (kWh)	Capacity charge (p kWh^{-1})
Rough	Depleted field	Space 66.5934	Space –
		Withdrawal 1 per day	Withdrawal 0.007
		Injection 0.3516	Injection 0.021
Hornsea	Salt cavity	Space 17.9487	Space –
		Withdrawal 1 per day	Withdrawal 0.008
		Injection 0.1108	Injection 0.024

Source: Storage Service Contract 2007; Rough and Hornsea Limited

obliged to publish (at least annually) terms and conditions for access to their facility (see Table 5.3). At this purpose, rights to use the capacity stored at each facility are sold for a standard storage year (i.e. 1st May–30th April). Any company is allowed to purchase storage capacity provided that the customer has signed the SSC and a credit agreement. Storage services are sold in Standard Bundled Units (SBUs).[9] And the capacity remaining unsold 30 days before the start of the new storage year must be offered for sale by auction and are awarded to the highest bidder on a pay-as-bid basis. Concerning the operation of Rough and Hornsea, the maximum storage capacity is offered to the market at all times and participation by customers is allowed by bilateral negotiations, auctions and/or other sales processes.

5.3.2 Security of Gas Supply

The debate in the UK on the need of implementing precautionary stocks is very interesting. Since 2004, it has been claimed that the fast liberalization process in the UK gas industry was overlooking security of supply (Stern, 2004; Wright, 2005). In fact, the increased dependency on gas has coincided with a decline in domestic production. In 2005 those patterns have led for the first time to a stress in UK supply-demand balancing. Security of supply concerns has entered the UK policy agenda during the winter of 2005–2006 on the wave of the continental climatic emergency. Since the UK gas demand was below expectations because of both mild-to-average weather (which shorten demand for heating) and high gas prices (which halved industrial requirements), in the winter of 2005–2006 it became clear that the tightness experienced by the UK gas market would have come entirely from a reduction in national supply. Several factors have contributed to that shortage. First, domestic gas production has declined more rapidly than envisioned by either the government or the industry. Second, gas import from the Continent via pipeline (Zeebrugge-Bacton interconnector) has been less than anticipated despite

[9] Differences in SBU are due to facility's technical parameters.

price differentials. Third, gas import via liquefied natural gas (LNG) terminals has resulted more idle than expected. Fourth, the sole viable seasonal storage facility (Rough) has been put off-line by a fire in February 2006.[10]

Over the last three years, the UK has thus moved from a position of relative self-sufficiency to one of import-dependence for gas. The greatest medium-term threat is the risk of disruption to gas supplies from mainland Europe; exposure to Russian gas in particular is viewed as a problem. On May 2006, the Secretary of State for Trade and Industry issued a "Statement on Need for Additional Gas Supply Infrastructure", recognizing a clear national need for new gas storage infrastructure. But on May 2007, the DTI response to the "Consultation on the effectiveness of current gas security of supply arrangements" pointed out a quite different attitude. Although regulating the use of storage by creating precautionary reserves could have a potentially beneficial impact on reducing the risk of involuntary interruptions in the short-term, it could significantly depress the revenue storage facilities can earn and thus lead to less investment in storage in the long-term. Moreover, DTI argues that with new storage and liquefied natural gas infrastructure being built, the system should be flexible enough to respond to unexpected interruption.

5.4 Precautionary Gas Stocks: A Model

The economy starts at date 0 in a state of abundance A and passes irreversibly at a random date in a state of crisis C. Time is continuous. The probability that the economy switches from A to C in a time interval dt is λdt, where λ is the publicly known parameter of this survival process. Thus, if the economy is in state A at some date, the economy will still be this state t periods later with probability $e^{-\lambda t}$. This simple modeling has three properties: (1) irreversibility ; (2) the crisis is certain only when it happens (no warning); (3) λ is independent of the state of inventories.[11] In any case, this structure represents the notion of low probability/high impact event (Stern, 2004).

We assume that consumers and producers only respond to the current price and the state $\sigma = A, C$. These responses are summarized by the "excess supply functions" $\Delta_\sigma[\cdot]$ defined over $\mathbb{R}_+^*$, where $\Delta_\sigma[p]$ is the difference in state σ and for price p between current *primary production* and current *final consumption*. For example,

[10] Comments from Ofgem: November 2005 "Gas storage was used heavily because beach (UK and Norwegian offshore gas) supplies, Interconnector and Isle of Grain were not delivering the expected amount of gas"; February 2006 "Rough gas storage facility was closed following a fire on February. The market responded with increasing supplies, particularly from the Interconnector".

[11] Since disruption risk linked with terrorist attack, civil war or pipeline breakdown can reasonably be seen as independent of accumulated reserves. Teisberg (1981) considers the deterrence effect of having sufficient reserves. However, the specification is given *a priori* and not founded on an explicit game between producer countries and the US.

$\Delta_C[\cdot]$ incorporates the supply shock and the adaptation of demand to the new state (e.g. use of interruptible contracts, fuel switching).[12]

Excess supply function $\Delta_\sigma[\cdot]$ is increasing and has a unique finite positive zero in R_+^*, denoted by p_σ^*; this is the price at which the spot market would be balanced without recourse to storage. Therefore, if the current price p is above p_σ^*, then the economy stores ($\Delta_\sigma[p] > 0$); if p is below p_σ^*, then the economy draws on gas inventories ($\Delta_\sigma[p] < 0$). Naturally, we assume that the abundance static equilibrium price p_A^* is strictly smaller than the crisis static equilibrium price p_C^*. See Fig. 5.2 for an illustrative example.

$\Delta_\sigma[\cdot]$ is a flow in the sense that if price p is sustained for the interval dt, then the quantity that is stored is $\Delta_\sigma[p]dt$. Thus, if we denote the total inventories in the economy by $S \geq 0$, conservation of matter imposes the following conditions

$$\begin{cases} \frac{dS}{dt} = \Delta_\sigma[p] & \text{if } S > 0 \text{ or } \Delta_\sigma[p] > 0, \\ \frac{dS}{dt} = 0 & \text{if } S = 0 \text{ and } \Delta_\sigma[p] \leq 0. \end{cases} \tag{5.1}$$

Storers are assumed to be risk-neutral price-takers, so that the price dynamics will be driven by arbitrage. Storage exhibits constant returns to scale. Carrying costs

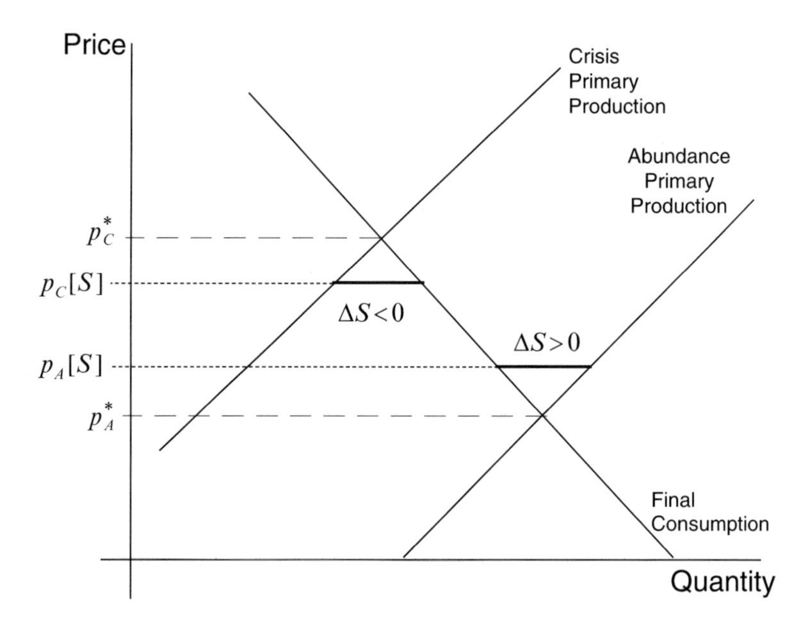

Fig. 5.2 Supply disruption and storage in the linear model

[12] This modeling is rationalizable with agents maximizing intertemporal utility or profit, provided objectives are time separable and quasi-linear. For a full justification, see Appendix 5.9.3 where surpluses are calculated.

consist of the opportunity cost of capital (r being the interest rate) and a cost c (per unit of commodity and per unit of time).[13]

We define the equilibrium as follows.

Definition 1. A competitive equilibrium starts at date 0, in state A, with some initial stocks S_0; it consists of contingent prices and stocks trajectories

$$\{p_A[t], p_C[t,\tau]\}_{t\geq 0,\tau\geq 0} \text{ and } \{S_A[t], S_C[t,\tau]\}_{t\geq 0,\tau\geq 0} \tag{5.2}$$

where t is the current date and τ the (random) date at which the crisis breaks out.

Three conditions must hold: (1) price-taking behavior by all agents (consumers, producers, storers); (2) rational expectations; (3) conservation of matter.

The date of the crisis τ has no impact on $p_A[\cdot]$ nor $S_A[\cdot]$; moreover $p_C[\cdot]$ and $S_C[\cdot]$ are defined only for dates posterior to the disruption. Non-strategic behavior of the agents, strictly increasing excess supply functions, linearity of the storage technology, risk-neutrality, all these hypotheses suffice to ensure that the competitive equilibrium is Pareto optimal.

5.5 Price and Stock Dynamics

Storers keep a stock of gas if expected price gains balance storage and interest cost. Whenever storages are non-empty, for a time increment dt, the no-arbitrage equations read

$$p_C[t,\tau] + c\,dt = (1 - r\,dt)\,p_C[t+dt,\tau], \qquad t \geq \tau, \tag{5.3}$$
$$p_A[t] + c\,dt = (1 - r\,dt)\left((1-\lambda dt)\,p_A[t+dt] + (\lambda dt)\,p_C[t+dt,t]\right). \tag{5.4}$$

In the above equations, the LHS is the unit price plus stockholding cost in states of crisis C and abundance A respectively. The RHS is the expected present unit value of the stocks after dt has elapsed. Equation (5.4) incorporates the risk of a regime switch. After elimination of second order terms, we get

$$\frac{\partial p_C[t,\tau]}{\partial t} = r p_C[t,\tau] + c, \qquad t \geq \tau, \tag{5.5}$$
$$\frac{dp_A[t]}{dt} = (r+\lambda)\,p_A[t] - \lambda p_C[t,t] + c. \tag{5.6}$$

We solve the model backwards. Once the crisis has broken out, the economy follows the Hotelling (competitive) dynamics; the gas price increases and the stocks shrink. Equation (5.5) is integrated for a fixed date τ and gives for all $t \geq \tau$:

---

[13] A more general structure with injection and withdrawal costs and limited storage capacity is discussed in Appendix 5.9.4.

$$p_C[t,\tau] = \min\left\{(p_C[\tau,\tau] + \frac{c}{r})\exp[r(t-\tau)] - \frac{c}{r}, p_C^*\right\}. \tag{5.7}$$

The price stops at p_C^* when the precautionary reserves are exhausted. Indeed, if the price were to overpass p_C^*, the economy would start accumulating gas without bound or time limit, which cannot be an equilibrium.

The economy drains the stocks that were in place at date τ, thus conservation of matter implies:

$$S_C[t,\tau] = -\int_t^{+\infty} \Delta_C[p_C[s,\tau]]ds, \tag{5.8}$$

$$S_A[t] = S_0 + \int_0^t \Delta_A[p_A[s]]ds, \tag{5.9}$$

$$S_A[t] = S_C[t,t] \text{ for all } t. \tag{5.10}$$

None of the model's parameters – interest rate, costs, crisis probability – depend on time. This simplifies the representation of the equilibrium, as the following proposition shows.

Proposition 6. *The equilibrium prices are only functions of current stocks. Functions $p_A[S]$ and $p_C[S]$ are continuous and decreasing for all $S \geq 0$; $p_C[S]$ has a simple implicit expression*

$$S = -\int_{p_C[S]}^{p_C^*} \frac{\Delta_C[p]}{rp+c}dp. \tag{5.11}$$

By using the results of Proposition 6 and (5.7), we obtain drainage duration for stocks S:

$$D[S] = \frac{1}{r}\ln\left[\frac{rp_C^*+c}{rp_C[S]+c}\right]. \tag{5.12}$$

This confirms that larger stocks always need more time to be drained. Drainage duration is necessarily finite: once the price has reached p_C^*, it would be uneconomical to keep costly stocks whose value will never increase.

The following proposition contains the fundamental properties of the equilibrium trajectories.

Proposition 7.

1. *The maximum inventories during abundance S^* is*

$$S^* = -\int_{\overline{p}_C}^{p_C^*} \frac{\Delta_C[p]}{rp+c}dp, \tag{5.13}$$

where

$$\overline{p}_C \equiv \left(\frac{r+\lambda}{\lambda}\right)p_A^* + \frac{c}{\lambda}. \tag{5.14}$$

S^* is positive if and only if $p_C^* > \overline{p}_C$. Moreover, S^* verifies $p_C[S^*] = \overline{p}_C$ and $p_A[S^*] = p_A^*$.

2. *The protection offered to the economy by the stocks has a maximum duration*

$$D^* = D[S^*] = \frac{1}{r} \ln \left[\frac{\lambda}{r+\lambda} \frac{rp_C^* + c}{rp_A^* + c} \right]. \tag{5.15}$$

3. *When $S^* > 0$, the economy approaches S^* without reaching it.*

The price threshold and the limit stocks are remarkably useful to describe the behavior of the economy. During the state of abundance, storers are willing to pay a premium proportional to the expected capital gains. As stocks approach S^*, these gains are progressively eroded and storers relax their pressure on prices. Accumulation slows down so much that the limit stock is never attained.

The time length D^* is positive if and only if S^* is positive. Maximum duration of drainage in (5.15) only depends on the boundary prices p_C^* and p_A^*, the interest rate and the unit cost. As a purely illustrative example, let's take c negligible with respect to the opportunity cost of the stock (price times interest rate). Limit stock and drainage time are non null if:

$$\frac{p_C^*}{p_A^*} > \frac{r+\lambda}{\lambda}. \tag{5.16}$$

For instance, with an interest rate of 5% and a "one-in-twenty-years" crisis ($\lambda = 5\%$ approximately), (5.16) implies that some precautionary storage takes place if the ratio p_C^*/p_A^* is larger than 2.

The impacts of parameters c, r, λ are unambiguous. With a higher unit storage cost or interest rate, the integrand in (5.13) decreases (the denominator increases) and the lower bound of integration $\overline{p}_C$ increases, thus S^* decreases. With a higher crisis probability, $\overline{p}_C$ is smaller, which gives a larger S^*. The effects on D^* are similar.

5.6 Dynamic Welfare Costs of Antispeculative Policy

In theory, governments should not interfere with security of supply, as competitive markets realize efficient solutions (Bohi and Toman, 1996). However, the Government might pursue short term political goals, supported by the consumers' pressure groups demanding stable supply of energy at an *affordable* price, no matter what the circumstances are (Mulder and Zwart, 2006). In view of this, storers would anticipate strict price controls.[14] Given the discouraging effects of this threat, the

[14] As Wright and Williams (1982) put it: "the oil industry has abundant reason to believe that there is some oil price at which Government will intervene to control the realizations of oil drawn down from private storage in times of shortage, when profit-maximizing private storers and importers

Government may wish to mitigate in advance its own foreseeable antispeculative intervention.[15] Our objective is to quantify the welfare loss of such second best policies.

The result of this political process that lead to specific strategic gas reserves can be summarized in terms of our model as follows. The policy consists of an "antispeculative" gas reserve price p_C^R which is smaller than p_C^* and independent of S for clarity. It is the price at which gas is sold and purchased as long as there are stocks to be drained. Since p_C^R induces a fixed drainage rate $\Delta_C[p_C^R]$, the price p_C^R is guaranteed for a fixed period only. From then on, stocks stay empty and the price is p_C^*. Since storing during crisis yields negative returns (the price cap prevents capital gains), storers sell all they have as soon as the crisis starts. To accommodate this, the Government can establish a public stabilization fund, which may either directly manage storage, or remunerate owners of storage facilities for their services, or pay stockholders their opportunity cost. All these schemes are equivalent as they engender the same surplus in total, though they differ as for how it is distributed across actors.

There are two cases, leading to very different equilibrium outcomes. If the crisis controlled price p_C^R is expected to be below $\overline{p}_C$, the smallest price that makes stockholding profitable, storage is totally discouraged in the abundance phase. If, on the contrary, p_C^R is above $\overline{p}_C$, storers see it as a price floor and they will not stop accumulation on their own in the abundance phase. Any inventory level can be attained if the crisis occurrence lags. To avoid this distortion, the Government has to put an upper bound on gas inventories, denoted by R. Here two variations are possible: either the abundance price is endogenous or it is also controlled by the Government.

We take the second option. Indeed, if p_A^R were determined by the market, arbitrage would make it equal to $\frac{\lambda}{r+\lambda} p_C^R - \frac{c}{r+\lambda}$ all along the accumulation phase. The stabilization fund established by the Government can replicate this price, hence our approach may be deemed rather general. Moreover, the theory of the second best says that p_C^R being distorted by political pressure, p_A^R may be voluntarily distorted by the Government: along with R, p_A^R serves to mitigate post crisis inefficiencies generated by the price cap.

To evaluate the antispeculative policy, we calculate the expected present surplus based on generated price and stocks trajectories. This yields a function of S, the stocks at the date the value is computed. Welfare being determined up to some arbitrary constant, we normalize our comparisons by setting at zero the value of the counterfactual no-storage policy (as if storage were impossible or too costly).

We denote the value of the optimal policy by $V_A^*[S]$ and the value of the antispeculative policy by $V_A^R[S]$. The following index measures welfare performance:

---

may well be branded as "speculators" or "price gougers". In fact, it may well be impossible for any administration credibly to guarantee against such action by itself or its successors."

[15] An alternative view is the following, described, for trade policy, in the lobby model of Grossman and Helpman (1994). Storers, requiring protection of their industry interest, can make implicit offers to the Government. The Government maximizes the sum of voters' welfare and total contributions from storers. A full-fledged version of this approach is beyond our scope.

$$v = \frac{V_A^R[S]}{V_A^*[S]}. \tag{5.17}$$

The maximum possible index is 1. A negative v would indicate a clear failure as the evaluated policy would do worse that no storage at all: the policy spoils resources by, e.g., building exaggerated stocks too fast and by using them too timidly. Such examples are (unfortunately for the society) quite easy to find as we shall see.

The detailed calculations of the total expected present surplus and of the index in (5.17) are relegated to Appendix 5.9.3.

5.7 To Build or not to Build Strategic Gas Stocks in the UK?

We now apply our model to evaluate the potential benefits of precautionary gas stocks in the UK. The application assumes linear excess supply functions:

$$\Delta_C[p_C] = bp_C - a; \quad \Delta_A[p_A] = \beta p_A - \alpha. \tag{5.18}$$

The reference prices are $p_C^* = a/b > p_A^* = \alpha/\beta$. Recall that Fig. 5.2 illustrates the supply disruption in the linear case.

We compare now four scenarios:

1. Competitive/surplus maximizing scenario;
   The three scenarios below are examples of the antispeculative type described before, i.e. summarizable by controls R, p_A^R and p_C^R:
2. Optimal antispeculative policy where all controls are optimized (they are denoted R^*, p_A^{R*} and p_C^{R*});
3. Antispeculative policies for various exogenous values of R, where only p_A^R and p_C^R are optimized;
4. A "test" (deliberately inefficient) policy characterized by excess reserves and exceedingly conservative management.

In scenario 1, the surplus maximizing limit stock S^* and drainage time D^* have explicit formulas that are calculated by using (5.13) and (5.15) respectively.[16] Moreover, $p_C[S]$ can also be calculated, whereas $p_A[S]$ and $V_A^*[S]$ are solved numerically.

We take parameters as roughly calibrated on the 2006 UK gas market as given in Table 5.4.

Time unit is the year, prices are in £/therm (1 therm = 2.76 m^3), quantities are expressed in billion therm. As for the probability of crisis ($\lambda = .02$), we consider

[16] The limit stock is

$$S^* = \frac{bc + ar}{r^2} \ln\left[\frac{\lambda}{r+\lambda}\frac{rp_C^* + c}{rp_A^* + c}\right] + \frac{b}{r}\left(\left(\frac{r+\lambda}{\lambda}p_A^* + c\right) - p_C^*\right). \tag{5.19}$$

The expression for D^* involves Lambert's W function, the inverse of $f(w) = we^w$.

Table 5.4 Parameters

Excess supply in C	$b = 0.95$	$a = 11.48$	$p_C^* = 12$
Excess supply in A	$\beta = 0.95$	$\alpha = 0.57$	$p_A^* = 0.6$
Costs	$r = 0.035$	$c = 0.15$	
Crisis arrival	$\lambda = 0.02$		

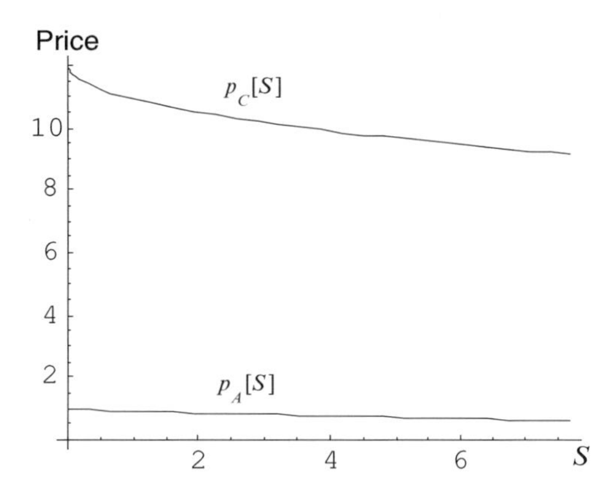

Fig. 5.3 Equilibrium price functions

the value that DTI (2006, p. 90) estimates as a "realistic chance of a significant supply interruption", based on ILEX (2006), JESS (2006), OXERA (2007) reports. The interest rate r and the maximal crisis price ($a/b = 12$) is taken from ILEX (2006, p. 106). This value, corresponding to an emergency cash out price, is assumed to reflect the damages to the economy of a sudden supply interruption. The average 2006 price ($\alpha/\beta = .6$) and annual consumption (about 36 billion therm) is documented by DTI (2007). The marginal cost of storage c is evaluated from available information, released by *Centrica Storage Ltd*, on the largest UK storage facility. Missing parameters are calculated with identifying assumptions: in case of major crisis, consumption could be reduced by 30% (price 12, inventories release notwithstanding). Finally we adopt a last (arbitrary) condition: $b = \beta$.

In Fig. 5.3, we show prices as a function of the stocks. Figure 5.4 depicts accumulation and drainage for alternative scenarios.[17] Accumulation starts at date $t = 0$ with $S = 0$ and the shock occurs at dates $t = 10, 20, \cdots, 80$. During the abundance phase, stocks are gradually piled up to approach $S^* = 7.7$ and the price decreases toward $p_A^* = 0.6$. When the crisis hits the economy, the price jumps to $p_C[S]$ and increases toward $p_C^* = 12$. Though it can take as long as $D^* = 5.4$ years, drainage appears to be much faster than accumulation.

As for the antispeculative policies, we numerically calculated the surplus they generate. Quite importantly, $V_A^R[0]$ can be expressed as an explicit function of constrained prices and target stock (R, p_A^R, p_C^R).

[17] Time unit is the year, prices are in £/therm and quantities in billion therm.

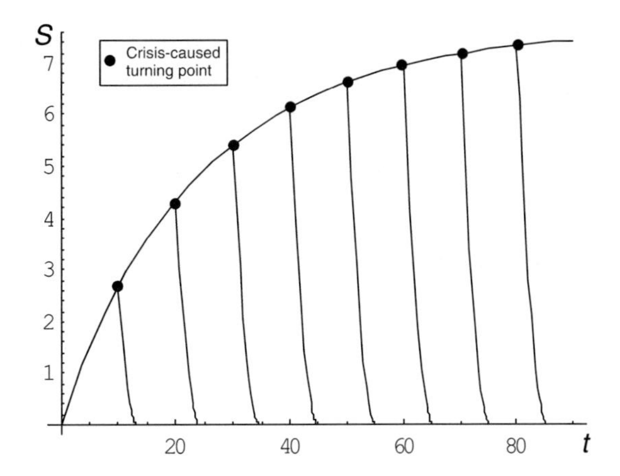

Fig. 5.4 Equilibrium stocks trajectories at various crisis dates

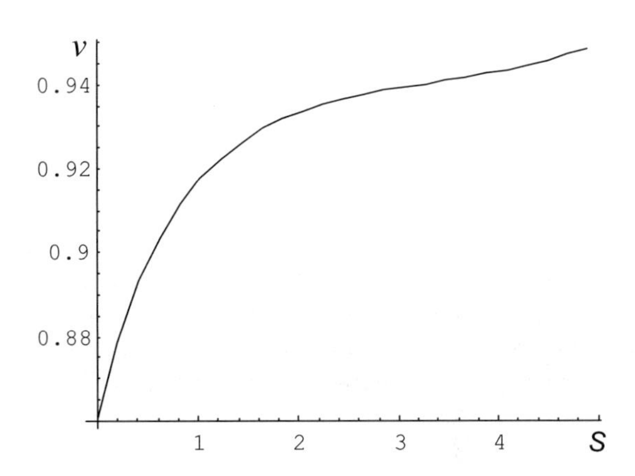

Fig. 5.5 Relative value of constrained optimal policies

For scenario 2, we found $R^* = 4.9$, $p_A^{R*} = 0.84$ and $p_C^{R*} = 10.5$. Accumulation takes 21.4 years, if no crisis breaks out before; drainage itself takes a maximum of 3.4 years.

Figure 5.5 displays the value v scenario 2 compared to that of scenario 1. Over the interval $[0, R^*]$ where both surpluses are defined, the index approaches 1 as inventories S increase:

- At $S = 0$, the suboptimal policy achieves 86% of the potential surplus;
- Gains increase very fast at the beginning of accumulation: at $S = 1$ (almost 20% of R^*), 64% of the initial efficiency loss are recouped;
- At R^*, 95% of the maximum surplus are captured by the suboptimal policy.

The latter effect is easily explained: as storage increases, the inefficiency of the *accumulation* strategy is sunk and thus disappears from the welfare comparison.

Let's turn to scenario 3. We calculated the performance of capacity constrained policies, i.e. for *given* values of R, as they could be imposed by physical availability, regulation or any other un-modeled decision process. For all tested values of R, forced (and constant) accumulation and withdrawal rates are optimized. We report the results graphically by showing the efficiency ratio of these policies as well as times required to implement the policy.

Figure 5.6 shows the efficiency index of all R in $[0, S^*]$. R being only a target, values are given at $S = 0$, namely, when the reserves program is initiated starting from scratch. We retrieve maximum efficiency attainable through such second-best policies for $R^* = 4.9$, which we already found in scenario 2. Small deviations from this optimum have of course second-order effects on efficiency. However, it appears, quite expectably, that the first units allocated to strategic storage have high social value. Dedicated reserves of the order of 2 Bscm would provide substantial benefits.

Figure 5.7 shows the *maximum* duration of the protection offered by the precautionary stocks for all values of R in $[0, S^*]$. It is calculated by dividing R by the optimal (constant) withdrawal rate. It is maximal in as much as, in reality, the

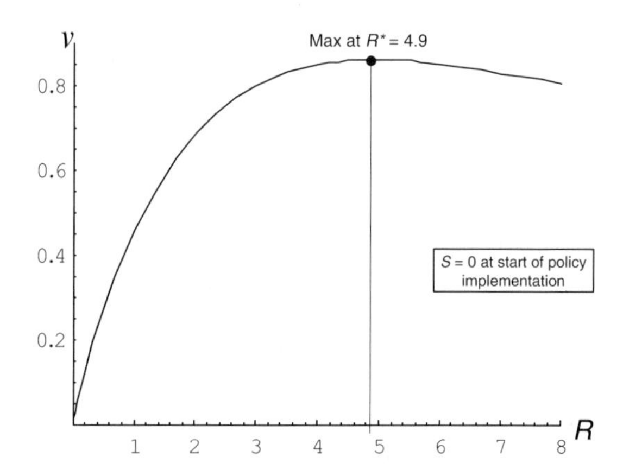

Fig. 5.6 Efficiency of second-best policy as a function of dedicated capacity (in Bscm). Policies are evaluated for empty reserves at the beginning

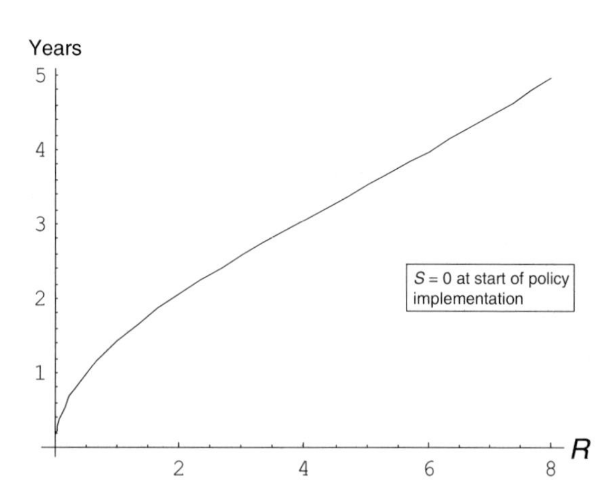

Fig. 5.7 Maximum duration of protection via second-best policy, as a function of dedicated capacity (in Bscm)

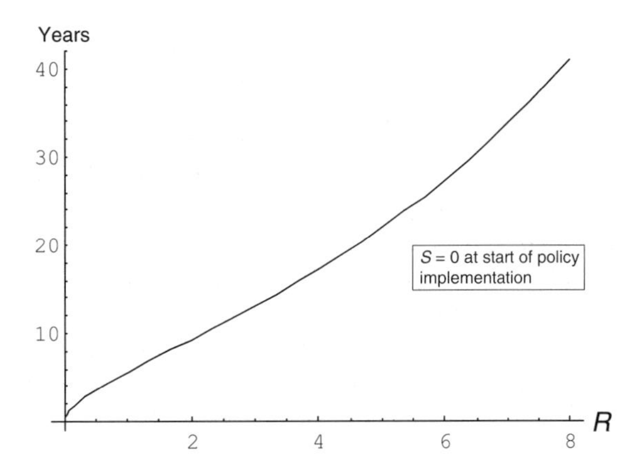

Fig. 5.8 Optimized time to build reserves via second-best policy, as a function of dedicated capacity (in Bscm)

crisis could occur before the capacity is completely filled. The marginal duration at $R = 0$ (no strategic reserves policy) is huge. Protection of the order of 9 months–1 year does not seem out of reach, as it suffices to set R at only 0.5 Bscm. The price smoothing effect would be moderate however, calling for more comfortable buffer stocks.

Figure 5.8 shows the time needed (if no crisis occurs before) to fill strategic reserves at the constrained optimal rate. The duration is given as a function of the target R. This illustrates that increasing R over 10–15 years to reach 2–3 Bscm seems to be an economically sensible policy.

Finally, we illustrate scenario 4, a deliberately inefficient policy. Though a no-reserves policy is costly, impatience and excess (too large reserves filled to fast) may be very detrimental to the economy. The expected present surplus is quite sensitive to the chosen policy. A simple example of a policy that dramatically under-performs the no-reserves option is proposed. Assume that the Government keeps $R = S^*$ (calculated from scenario 1) as a target but imposes a twice larger accumulation rate and a twice slower drainage rate than those obtained under scenario 2. In short, the capacity required is too large, the accumulation too fast and the withdrawal too slow. This policy can be seen as extremely prudent.

Figure 5.9 shows the relative value of this prudent policy as a function of the *current* stocks S, with $S \in [0, S^*]$.

At zero stocks and up to $S = 1.4$ approximately, the policy imposes huge welfare costs (the index starts at -0.72). This means that the economy would be better off if storage were impossible (or $R = 0$).

Due to fast accumulation, the price is very high during the accumulation phase, which penalizes consumers; in addition, the economy sustains the cost of excessive reserves. This effect becomes attenuated as storage expenditures get sunk, but to a much lesser extent than with the constrained optimum.

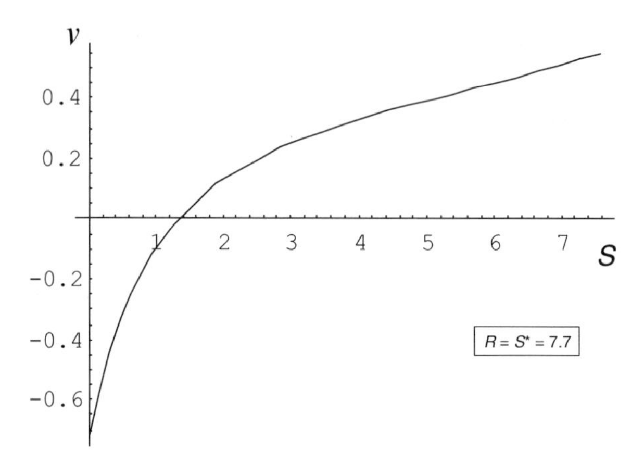

Fig. 5.9 Relative value of inefficient policy

5.8 Conclusion

We developed a model of optimal stockpiling and reserve duration to face up to a potential irreversible supply shock. Our key ingredient is that price trajectories, accumulation and drainage behavior are interdependent in equilibrium. This differentiates the approach from inventory management models in which prices are given, or precautionary reserve studies in which the welfare costs of building the stocks are ignored. A simple condition determines whether precautionary stocks should be accumulated. General cost structures, in particular limited storage capacity, are shown to have intuitive and calculable effects on the main properties of the equilibrium.

The calibration of our model on the UK data shows that policy interventions, tough motivated by antispeculative reasons, could create welfare losses. Our results suggest that the UK decision of not stockpiling precautionary gas stocks could be inefficient. In fact, no energy system is invulnerable to the possibility of an interruption to one or more supply sources or to fluctuations to demand levels. Our model shows that whether such shocks to the demand-supply balance lead to a "gas crisis" will depend on the interaction between the prevailing market situation (tight or well supplied) and the nature of the shocks (mainly the time at which they occur and the demand and supply response).

5.9 Appendix

5.9.1 Proof of Proposition 6

The RHS of (5.8) is strictly increasing in the value of $p_C[\tau,\tau]$, thus it gives a unique strictly decreasing relationship between p_C and $S \geq 0$, denoted by $p_C[S]$. We can then define $p_A[S]$ by backwards induction.

We can now replace the price dynamics in (5.5) and (5.6) by

$$\Delta_C[p_C[S]] \cdot \frac{dp_C[S]}{dS} = rp_C[S] + c, \tag{5.20}$$

$$\Delta_A[p_A[S]] \cdot \frac{dp_A[S]}{dS} = (r+\lambda)p_A[S] - \lambda p_C[S] + c, \tag{5.21}$$

for $S > 0$.

Equation (5.20) can be integrated directly to get (5.11). The RHS of (5.21) cannot be positive (otherwise storers would liquidate inventories at once) implying that $\frac{dp_A[S]}{dS} < 0$.

5.9.2 *Proof of Proposition 7*

1. Remark that $\overline{p}_C$ is the minimum value $p_C[\cdot]$ can take: storers are just indifferent between keeping or selling their stocks if the abundance price is as low as p_A^*, since the carrying costs ($rp_A^* + c$ per unit) equals the *expected* earning ($\lambda(p_C[S^*] - p_A^*)$ per unit). The corresponding stocks are denoted by S^*; S^* being the maximum stocks, it verifies $p_C[S^*] = \overline{p}_C$ and $p_A[S^*] = p_A^*$.

This reasoning implies in particular that if $\overline{p}_C \geq p_C^*$, then $S^* = 0$: holding inventories cannot be profitable and the crisis will simply cause a price jump from p_A^* to p_C^*.

2. By plugging $\overline{p}_C$ into (5.11), we obtain the expression in the text.

3. The price p_A must converge continuously towards p_A^* before the occurrence of the gas disruption. As p_A covers half its difference with the limit p_A^*, the variation rate of the stock per unit of time Δ_A is approximately halved (the derivative of excess demand at p_A^* is not zero), meaning that the convergence speed dS/dt is approximately halved. This implies that, whatever the proximity of the limit, the duration to cover half the distance to the limit is approximately constant, thus the limit is not attained in finite time.

5.9.3 *Expected Present Surplus*

Consider a representative consumer whose intertemporal utility function valorizes gas consumption and a separable numéraire. Leaving aside uncertainty at this stage, the consumer's objective can be written as

$$\int_0^{+\infty} (u_\sigma[q_t] - p_t q_t)e^{-rt}dt, \; \sigma = A, C, \tag{5.22}$$

where u_σ is a state dependent, increasing and concave utility, q_t is date t gas consumption and $p_t q_t$ is date t expenditure. Consider also a representative producer whose technology can by aggregated at t by a state dependent convex cost function $C_\sigma[q_t]$.

For a given price p, final demand is $u_\sigma'^{-1}[p]$ and primary production is $C_\sigma'^{-1}[p]$, thus excess supply functions as we defined them can be expressed

$$\Delta_\sigma[p] = C_\sigma'^{-1}[p] - u_\sigma'^{-1}[p]. \tag{5.23}$$

The instantaneous surplus depends only on the state σ, S and the current price p

$$W_\sigma^0 + W_\sigma[p] - cS \tag{5.24}$$

where W_σ^0 denotes the *reference surplus*, i.e. calculated at price p_σ^*, and where cS is the cost of keeping the inventories. The key point is that $W_\sigma[p]$ can be derived from the excess supply function $\Delta_\sigma[p]$:

$$W_\sigma[p] = \int_{p_\sigma^*}^{p} \Delta_\sigma[x]dx - p\Delta_\sigma[p], \tag{5.25}$$

as we can directly see in Figs. 5.10 and 5.11.

The final step consists of calculating the expected present surplus, by discounting all future instantaneous surpluses and then taking the expectation. Given the initial state A, stock S_0 at date 0 and the stockholding dynamics, the expected intertemporal surplus of the *optimal policy* is denoted:

$$V_A^0 + V_A^*[S_0], \tag{5.26}$$

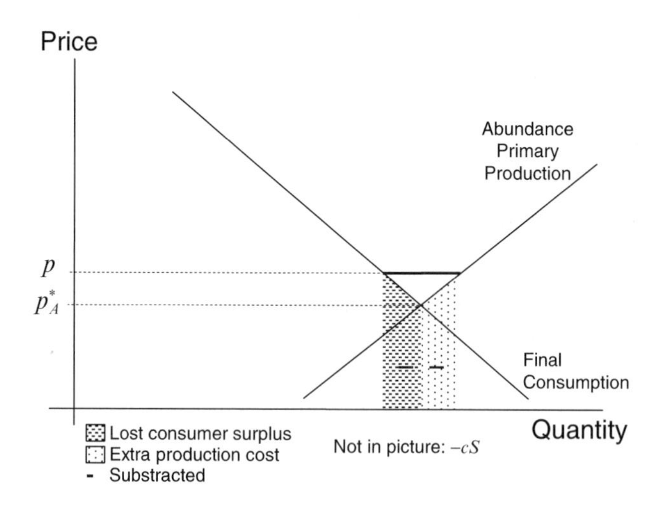

Fig. 5.10 Surplus during abundance

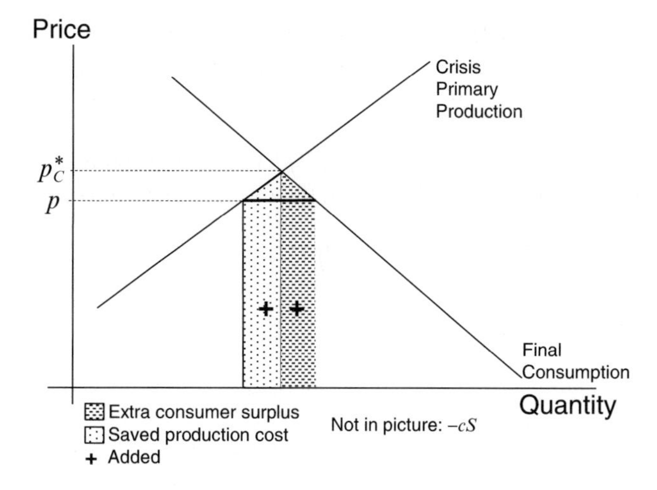

Fig. 5.11 Surplus during crisis

where V_A^0 is defined as

$$V_A^0 = \frac{W_A^0}{r+\lambda} + \frac{\lambda W_C^0}{r(r+\lambda)}, \tag{5.27}$$

and

$$V_A^*[S_0] = E\left\{ \int_0^{+\infty} \left(W_{\sigma_t}[p^*_{\sigma_t}[S_t]] - cS_t\right) e^{-rt} dt \right\}, \tag{5.28}$$

σ_t being the (random) state at date t.

The antispeculative policy (summarized by p^R_σ and constrained accumulation and drainage functions $\Delta_\sigma[p^R_\sigma]$) will generate the instantaneous surplus $W^*_\sigma + W^R_\sigma[p^R_\sigma] - cS$. Total expected present surplus is:

$$V_A^0 + V_A^R[S_0], \tag{5.29}$$

where V_A^0 is defined by (5.27) and

$$V_A^R[S_0] = E\left\{ \int_0^{+\infty} \left(W_{\sigma_t}[p^R_{\sigma_t}] - cS_t\right) e^{-rt} dt \right\}. \tag{5.30}$$

The Bernoulli process driving the evolution of σ_t being exogenous and time independent, the terms comprising W_A^0 and W_C^0 are identical whatever the policy evaluated and therefore V_A^0 can be normalized at zero. This is why we state that the no storage policy (a useful reference) can be given null value. The relative value of a given policy with respect to the optimum is therefore correctly captured by the index:

$$v = \frac{V_A^R[S]}{V_A^*[S]}. \tag{5.31}$$

5.9.4 Extensions of the Basic Model

Injection and Release Costs

The analysis can be easily extended to the case where the costs of injecting and releasing gas are non negligible. Denote unit injection cost by i and unit release cost by s. With perfect competition, gas *outside* and *inside* the reservoir state can be traded at prices that we denote respectively by $p_\sigma[S]$ and $p^I_\sigma[S]$ (with $\sigma = A, C$ and $S \geq 0$). The market equilibrium between outside and inside gases implies that, whenever $S > 0$,

$$p_A[S] + i = p^I_A[S] \text{ and } p_C[S] = p^I_C[S] + s. \quad (5.32)$$

The structure of the system of equations is preserved, with p^I_σ replacing p_σ. Arbitrage conditions (5.20) and (5.21) become

$$\Delta_C[p^I_C + s] \cdot \frac{dp^I_C}{dS} = rp^I_C + c, \quad (5.33)$$

$$\Delta_A[p^I_A - i] \cdot \frac{dp^I_A}{dS} = (r+\lambda)p^I_A - \lambda p^I_C + c. \quad (5.34)$$

Remark that the excess supply functions are shifted, thus boundary conditions are

$$p^I_C[0] = p^*_C - s, \quad (5.35)$$

$$p^I_A[S^*] = p^*_A + i. \quad (5.36)$$

The range of p^I_σ is narrower than that of p_σ: the minimum is higher, the maximum is lower. As a result, the condition ensuring positivity of the limit stock is more restrictive, i.e.

$$p^*_C - s > \left(\frac{r+\lambda}{\lambda}\right)(p^*_A + i) + \frac{c}{\lambda}. \quad (5.37)$$

Expressions of optimal limit stock and drainage time are now based on shifted excess supply functions and shifted boundary prices.

Limited Storage Capacity

Gas is mostly stored in depleted fields and aquifers; the development of such facilities is naturally limited. If the capacity devoted to precautionary storage K exceeds S^* previously calculated, then the unconstrained solution remains valid; otherwise, the maximum stock is constrained to equal K, which in turn affects price trajectories and the value of storage facilities.

During the crisis, $p_C[S]$ is unchanged compared to the unconstrained case. Reserves are gradually drained, meaning that the storage price, under competitive assumption, remains fixed at the marginal cost c. In the abundance state, the price function $p^K_A[S]$ depends on K: the accumulation process must stop when capacity is saturated, therefore $p_A^K[K] = p_A^*$. The storage price is also c as long as some capacity remains vacant; when K is attained, it jumps to $\pi_A^K > c$, with

$$\pi_A^K = \lambda(p_C[K] - p_A^*) - rp_A^*. \tag{5.38}$$

The net rent $\pi_A^K - c$, captured by the owners of the storage capacity, balances the carrying costs of a fixed stock with its expected benefits. Storage capacity units gain value as K diminishes. This combines two effects: the smaller K becomes, the larger π_A, and also the shorter the time before saturation will be.

The first effect (the monotonicity of π_A) derives directly from the monotonicity of $p_C[K]$. The second effect is shown as follows.

Monotonicity of the scarcity rent

The function p_A^K follows ODE (5.21), with boundary condition $p_A^K[K] = p_A^*$. As the function p_C is independent of K, the Cauchy–Lipschitz theorem implies that the price functions for two different capacities below S^* never cross. Thus for all $S \in [0,K]$ and $K < K'$, $p_A^K[S] < p_A^{K'}[S]$ with both functions decreasing. We now show that the time T_K needed for the price to pass from $p_A^K[0]$ to p_A^* is longer the larger is the capacity K. Using (5.21), we have

$$T_K = -\int_{p_A^*}^{p_A^K[0]} \frac{dp_A}{(r+\lambda)p_A - \lambda p_C[p_A^{K(-1)}[p_A]] + c}. \tag{5.39}$$

Given the monotonicity of p_A^K with respect to K, the above sum with a larger K integrates a function of higher absolute value over a longer interval. This gives us the announced result.

References

Bergstrom, C., Loury, G. C., and Persson, M. (1985). Embargo threats and the management of emergency reserves. *Journal of Political Economy, 93*(1), 26–42.

Bohi, D. R. and Toman, M. A. (1996). *The economics of energy security*. Berlin: Springer.

Crawford, V., Sobel P. J., and Takahashi, I. (1984). Bargaining, strategic reserves, and international trade in exhaustible resources. *American Journal of Agricultural Economics, 66*(4), 472–480.

Devarajan, S. and Weiner, R. J. (1989). Dynamic policy coordination: Stockpiling for energy security. *Journal of Environmental Economics and Management, 16*(1), 9–22.

DG TREN, (2006a). Statistical pocketbook 2006, Brussels.

DG TREN, (2006b). European Energy and Transport, Trends to 2030, Brussels.

DTI, (2006). The energy challenge. London.

DTI, (2007). 2007 Energy market outlook. London.

Directive 98/30/EC of the European Parliament and of the Council concerning common rules for the internal market in natural gas.

Directive 2003/55/EC of the European Parliament and of the Council concerning common rules for the internal market in natural gas.

Directive 2004/67/EC of the European Parliament and of the Council concerning measures to safeguard security of natural gas supply.
Gas Storage Database (2008).
Grossman, G. and Helpman, E. (1994). Protection for sale. *American Economic Review, 84*, 833–850.
Hillman, A. L. and Van Long, N. (1983). Pricing and depletion of an exhaustible resource when there is anticipation of trade disruption. *Quarterly Journal of Economics, 98*(2), 215–233.
Hogan, W. (1983). Oil stockpiling: Help thy neighbor. *Energy Journal, 4*(3), 49–71.
Hughes, H. A. J. (1984). Optimal stockpiling in a high-risk commodity market: The case of Copper. *Journal of Economic Dynamics and Control, 8*(2), 211–238.
ILEX (2006). Strategic Storage and Other Options to Ensure Long-Term Security of Supply. Report to DTI.
JESS (2006). Long-term security of energy supply. Report to DTI.
Mulder, M. and Zwart, G. (2006). Market failures and Government policies in gas markets. CPB Memoranda 143, CPB Netherlands Bureau for Economic Policy Analysis.
Nichols, A. and Zeckhauser, R. (1977). Stockpiling strategies and Cartel prices. *Bell Journal of Economics, 8*(1), 66–96.
OXERA, (2007). An assesment of potential measure to improve gas security of supply. Report to DTI.
Scottish and Southern Energy (2007). Storage services contract – The Hornsea storage facility.
Stern, J. (2004). UK gas security: Time to get serious. *Energy Policy, 32*(17), 1967–1979.
Stiglitz, J. (1977). An economic analysis of the conservation of depletable natural resources. Draft Report, IEA, Section III.
Sweeney, J. (1977). Economics of depletable resources: Market forces and intertemporal bias. *Review of Economic Studies, 44*, 125–142.
Teisberg, T. J. (1981). A dynamic programming model of the U.S. strategic petroleum reserve. *Bell Journal of Economics, 12*(2), 526–546.
Tolley, G. S. and Wilman, J. D. (1977). The Foreign dependence question. *Journal of Political Economy, 85*(2), 323–347.
Weisser, H. (2007). The security of gas supply – a critical issue for Europe? *Energy Policy, 35*(1), 351–355.
Williams, J. C. and Wright, B. D. (1991). *Storage and Commodity Markets*. Cambridge: Cambridge University Press.
Wright, B. D. and Williams, J. C. (1982). The roles of public and private storage in managing oil import disruptions. *Bell Journal of Economics, 13*, 341–353.
Wright, P. (2005). Liberalisation and the security of gas supply in the UK. *Energy Policy, 33*, 2272–2290.

Chapter 6
Final Remarks and Policy Recommendations

Monica Bonacina, Anna Cretì, and Antonio Sileo

In the last decade storage has become a priority for both gas companies and governments. Spurred by declining national gas reserves, inflexible production patterns and expensive alternatives to balance supply and demand in the yearly gas cycle, storage is vital to the effective and efficient functioning of national gas markets. The book focuses on Italy, France, Germany and the UK. These Countries have contributed to about 62% of demand for natural gas in 2007 in the EU-27 (72% with respect to the EU-15) and they manage more than 60% of the installed working capacity.[1] However, these Countries show several – even opposite – patterns of development.

Storage is the primary flexibility tool in import dependent countries, like Germany, Italy and France. Such pivotal role will be reinforced by a new wave of investments in gas storage facilities. Planned investments are particularly ambitious in the UK, notwithstanding the usage of production to accommodate seasonal swing and the small share of imports (see Table 6.1).

Storage services differ in terms of gas deliverability because of geological constraints. Italy only uses depleted reservoirs; Germany benefits from all type of gas storage technologies; UK and France have a good mix of seasonal and peak shaving facilities.

Storage usage reflects different consumption patterns. The difference between winter and summer consumption is very high in France, where demand is quite inelastic, gas being consumed mainly by residential customers. In Italy, Germany and UK, instead, consumption is sustained by the demand for electric generation,

M. Bonacina and A. Sileo
IEFE (Centre of Research on Energy and Environmental Economics and Policy), Bocconi University, via Roentgen, 1, 20136 Milan, Italy
e-mail: {monica.bonacina, antonio.sileo}@unibocconi.it

A. Cretì
IEFE (Centre of Research on Energy and Environmental Economics and Policy), and Bocconi University, Department of Economics, via Roentgen, 1, 20136 Milan, Italy
e-mail: anna.creti@unibocconi.it

[1] Our estimates on data by EUROSTAT.

A. Cretì (ed.), *The Economics of Natural Gas Storage: A European Perspective*,
DOI: 10.1007/978-3-540-79407-3_6,

Table 6.1 Stock coverage and planned investments

Country	Stock change/consumption (Oct. to March – %)	Investments by 2012 (Bscm)
Germany	11.6	1.36
France	21.4	0.34
Italy	9.9	3.43
UK	3.9	7.86

Source: our elaboration on EUROSTAT and GSE data, 2007–08

Table 6.2 Gas cycle

Country	Summer consumption April to Sept. (Bscm)	Winter consumption Oct. to March (Bscm)
Germany	34.7	60
France	12.7	35.5
Italy	32.3	55.2
UK	38.2	66.4

Source: our elaboration on EUROSTAT data, 2007–08

Table 6.3 Regulation. Source: our elaboration on country level information, 2007

Country	Storage Oper. (SOs)	Market Concentr.*	TPA access	TPA tariffs	Storage for Supply Sec.
Germany	>20	26	N	SOs	No
France	2	79	N	SOs	PSOs
Italy	2	98	R	Regulated	Yes
UK	5	75	N/R	SOs/Regulated	No

* Refers to % of capacity by the biggest Storage Operator

which is smoother over the year and more elastic than demand for home heating (see Table 6.2).

The organization of storage services also depends on the extent of gas market competition and regulatory options. As Table 6.3 shows, storage capacity is often in the hand of incumbents, as for instance in France and Italy, but can also be more fragmented, as in Germany, or unbundled, as in the UK. Light-hand access regulation is a common feature to Germany and France, with the polar cases of Italy on one side, where access is set at cost-reflective tariffs, and the UK, on the other, where a market-oriented approach is developed. Storage for security of supply is a constraint directly imposed in Italy by the Ministry of Industry (with a strategic reserve amounting to 38% of the system working capacity), whereas in France rules for allocation of capacity create Public Service Obligations (PSOs), despite negotiated access. Germany and UK do not have gas immobilized for security purposes.

The variety of the observed patterns in Germany, France, Italy and the UK is representative, we believe, of relevant future trends in the storage sector across Europe.[2] The heterogeneity of these countries is taken into account by four specific models, each Chapter focusing on crucial aspects of the economics of storage. Competition policy issues, regulation and security of supply are the topics analyzed.

To summarize the main message of the "four stories" in a nutshell, we could say that despite dissimilar features in the gas market at the country level, the less storage is market oriented, the more crucial the role of regulation. Welfare costs due not only to imperfect competition, but also to imperfect policies are not negligible.

Chapter 2 gives empirical evidence to a *fundamental problem that prevents the efficient use storage facilities, i.e. the lack of adequate market signals that provide arbitrage opportunities.* This is relevant not only for Germany, the country analyzed in this Chapter, but also for the South-South East zone, where a proper standardized exchange does not exist.[3] Other possible reasons to explain imperfect intertemporal arbitrage are technical constraints (to provide short-run balance of supply and demand or to regulate the pipeline pressure), the role of storage facilities for security of supply purposes, cost of storage operations and, most importantly, exercise of market power. These arguments pave the way to a critical question, that is the optimal regulation of storage systems. As long as it is recognized that storage is not a natural monopoly (as well argued in Chap. 3), specific rules should apply to that market. In fact, dominant positions both in the horizontal dimension (Chap. 3) and in the vertical structure (Chap. 4) negatively affect the competitive organization of storage services.

The value of access charges to storage facilities might distort decisions on downstream market and gas sourcing. Imperfect regulation in the form of a low access price leads the strategic use of storage, whenever alternative flexibility instruments are costly. Storage helps gas firms to create endogenous leadership using little flexible facilities, according to Baranès, Mirabel and Poudou. Rivals may be pushed to raise their sourcing cost using both spot market and more flexible storage facilities. Vertical integration destroys those negative externalities thus resulting in higher welfare. A trade-off appears: low access prices encourage preemptive access to storage, but vertical economies offset costs of lower gas availability. *For policy makers, this is an argument against unbundling, which is in sharp contrast with actual EU proposals.*

Cavaliere comes back to the problem of limited availability of flexibility instruments together with the lack of storage capacity. Low storage tariffs are shown to

[2] Each of these countries belongs to one of the four Regional Energy Markets (REM) identified by the European Regulators' Group for Electricity and Gas as the most important zonal gas markets that will develop in next years. The REM are: North West (Netherlands, Belgium, France, UK, Ireland), North (Germany, Netherlands, Sweden, REM) South (Spain, Portugal and Southern France) and South-South East (Italy, Austria, Slovakia, Hungary, Slovenia, Greece, Poland and the Czech Republic).

[3] This zone is both an important market in itself with a consumption of 140 billion cubic meters per year (28% of the European Union), and a key transit area (nearly 100% of Russian imports to Europe is transported through it).

create a drawback as they might not signal scarcity. This effect is magnified with imperfect competition in the final gas market. When a leader controls the final gas price, he has the incentive to increase his storage services demand to exclude the follower from the final market. Administrative rules and long-term booking of storage capacity are detrimental to competition, as they allow the dominant firm to carry out this kind of exclusionary strategies. Market mechanisms as auction of capacity are welfare improving only if coupled to relatively high access prices. *In the regulatory perspective, this is an argument against low access prices.*

Cretì and Villeneuve contribute to the ongoing debate on security of supply in competitive markets. At the EU policy level, a contradiction seems to arise. The 2007 proposal for a new liberalization package considers that building up strategic gas stocks to deal with potential supply disruptions would be too expensive and technically difficult at the moment. Yet a few years ago, the message delivered by the Green Paper on Energy Security of Supply and the Directive 2004/67/EC was different, encouraging Member States to take specific measures against supply shocks. The dynamic stochastic model developed in Chap. 5 focuses on low probability-high impact events, thus characterizing the nature of eventual gas supply disruption at the European level. The probability of price increase after a gas crisis provides enough incentives to private stockholders who accumulate precautionary reserves. This is in line with the EU wisdom that competition and security of supply go hand in hand. However, given the unavoidable policy dimension of gas security, government rules might interfere with decentralized stockholding decisions. The model's simulations, calibrated on the 2006 UK gas market, point out the danger of such imperfect policies. *Administrative rules that prevent price spikes have non negligible costs, but the "do nothing strategy" does not perform better. This conclusion should alert those countries who are relaxing their attention on precautionary gas stocks.*

In the near future, a European market for storage will not be achieved as long as a truly European liberalized gas market is not organized. To this end, new infrastructure such as liquefied gas terminals and pipelines, supply diversification strategies, creation of liquid spot markets will play a crucial role. Storage will follow.

Printing: Krips bv, Meppel, The Netherlands
Binding: Stürtz, Würzburg, Germany